T0214284

Basics of Probability and Stochastic Processes

Esra Bas

Basics of Probability
and Stochastic Processes

 Springer

Esra Bas
Industrial Engineering
Istanbul Technical University
Istanbul, Turkey

ISBN 978-3-030-32325-7   ISBN 978-3-030-32323-3   (eBook)
https://doi.org/10.1007/978-3-030-32323-3

This Springer imprint is published by the registered company Springer Nature Switzerland AG
The registered company address is: Gewerbestrasse 11, 6330 Cham, Switzerland

Preface

This book is aimed as a textbook for one-semester course in Introduction to Probability and Stochastic Processes to be taught at engineering schools at the undergraduate level. Since I teach this topic to students with different engineering majors, I thought a book at a very basic level without theoretical details would be beneficial. It is organized so that the students with no prior knowledge can learn about the basic concepts of probability and stochastic processes in a step-by-step manner and get insights by reading numerous remarks and warnings. All the examples and problems are solved step-by-step by assuming that the students have only basic calculus knowledge. The chapters include some basic examples, which are revisited as a new concept is introduced. Since I believe that engineering students can acquire knowledge by visual means more easily, I added several figures and diagrams to facilitate the comprehension of the basic concepts and the solutions of the examples and problems. I also used a table format where relevant so that the concepts and formulae can be understood in comparison with each other. This table format is also intended to serve as a summary of crucial formulae. I tried to keep each chapter simple with a few sub-chapters and independent from the other chapters.

This book has two main parts. In the first main part of this book, the readers can get familiar with the basics of probability including combinatorial analysis, conditional probability, discrete and continuous random variables, and other selected topics in probability including jointly distributed random variables, while in the second main part of this book, they learn the basics of stochastic processes including point process, counting process, renewal process, regenerative process, Poisson process, Markov chains, queueing models and reliability theory. The topics are presented from broad to detailed levels. As an example, Chap. 4 is devoted to "Introduction to Random Variables", which provides the basics that are relevant to both discrete and continuous random variables. However, the readers can learn more about "Discrete Random Variables" and "Continuous Random Variables" in Chaps. 5 and 6, respectively. As another example, Chap. 9 is devoted to "A Brief Introduction to Point Process, Counting Process, Renewal Process, Regenerative Process, Poisson Process". In this chapter, the readers can understand the basic relations between these basic subjects. Afterwards, the interested readers can also learn more about Poisson Process and Renewal Process in Chaps. 10 and 11, respectively. I deliberately included only basic concepts in each chapter. As an example, only birth and death queueing models are provided in Chap. 15, since the emphasis of this book is on the concepts at a very basic level.

Although the primary audience of this book is all engineering students at the undergraduate level, the graduate students who want to refresh their basic knowledge about probability and stochastic processes can also use this book as a review before starting with the course stochastic processes at the graduate level. Since I have also been teaching this topic at the graduate level, the importance of recalling the basic concepts has been very clear to me based on my teaching experiences.

Istanbul, Turkey   Esra Bas

Contents

Part I Basics of Probability

1 Combinatorial Analysis 3
   1.1   The Basic Principle of Counting and the Generalized Basic
         Principle of Counting 3
   1.2   Combinatorial Analysis (Combinatorics) 5

2 Basic Concepts, Axioms and Operations in Probability 15
   2.1   Basic Concepts 15
   2.2   Axioms of Probability 16
   2.3   Basic Operations in Set Theory Versus Basic Operations
         in Probability Theory 17

3 Conditional Probability, Bayes' Formula, Independent Events 27
   3.1   Conditional Probability 27
   3.2   Bayes' Formula 30
   3.3   Independent Events 32

4 Introduction to Random Variables 39
   4.1   Random Variables 39
   4.2   Basic Parameters for the Discrete and Continuous Random
         Variables 40

5 Discrete Random Variables 55
   5.1   Special Discrete Random Variables 55
   5.2   Basic Parameters for Special Discrete Random Variables 59

6 Continuous Random Variables 71
   6.1   Special Continuous Random Variables 71
   6.2   Basic Parameters for Special Continuous Random Variables ... 83

7   Other Selected Topics in Basic Probability 95
    7.1   Jointly Distributed Random Variables 96
    7.2   Conditional Distribution, Conditional Expected Value,
         Conditional Variance, Expected Value by Conditioning,
         Variance by Conditioning 105
    7.3   Moment Generating Function and Characteristic Function 108
    7.4   Limit Theorems in Probability 110

Part II   Basics of Stochastic Processes

8   A Brief Introduction to Stochastic Processes 125

**9   A Brief Introduction to Point Process, Counting Process,
Renewal Process, Regenerative Process, Poisson Process** 131
    9.1   Point Process, Counting Process 132
    9.2   Renewal Process, Regenerative Process 134
    9.3   Poisson Process 135

10   Poisson Process 149
    10.1   Homogeneous Versus Nonhomogeneous Poisson Process 149
    10.2   Additional Properties of a Homogeneous Poisson Process 153

11   Renewal Process 163
    11.1   Basic Terms 163
    11.2   Limit Theorem, Elementary Renewal Theorem,
          Renewal Reward Process 165
    11.3   Regenerative Process 170

12   An Introduction to Markov Chains 179
    12.1   Basic Concepts 179
    12.2   Transition Probability Matrix, Matrix of Transition Probability
          Functions, Chapman-Kolmogorov Equations 181
    12.3   Communication Classes, Irreducible Markov Chain,
          Recurrent Versus Transient States, Period of a State,
          Ergodic State 183
    12.4   Limiting Probability of a State of a Markov Chain 189

13   Special Discrete-Time Markov Chains 199
    13.1   Random Walk 199
    13.2   Branching Process 202
    13.3   Hidden Markov Chains 204
    13.4   Time-Reversible Discrete-Time Markov Chains 205
    13.5   Markov Decision Process 207

14   Continuous-Time Markov Chains 217
    14.1   Continuous-Time Markov Chain and Birth & Death Process ... 218
    14.2   Birth & Death Queueing Models 221

14.3 Kolmogorov's Backward/Forward Equations, Infinitesimal
Generator Matrix of a *CTMC* and Time-Reversibility
of a *CTMC*.. 224

15 An Introduction to Queueing Models 233
15.1 Basic Definitions 233
15.2 Balance Equations and Little's Law Equations for *B&D*
Queueing Models 235

16 Introduction to Brownian Motion......................... 253
16.1 Basic Properties of a Brownian Motion 253
16.2 Other Properties of a Brownian Motion 256

17 Basics of Martingales 265
17.1 Martingale, Submartingale, Supermartingale,
Doob Type Martingale 265
17.2 Azuma-Hoeffding Inequality, Kolmogorov's Inequality,
the Martingale Convergence Theorem 269

18 Basics of Reliability Theory 273
18.1 Basic Definitions for a Nonrepairable Item................ 274
18.2 Basic Definitions for a Repairable Item 278
18.3 Systems with Independent Components 280
18.4 Systems with Dependent Components 284

Area Under the Standard Normal Curve to the Left of *z* 293

References 297

Index 299

Part I
Basics of Probability

Chapter 1
Combinatorial Analysis

Abstract In this chapter, the basics of combinatorial analysis including the basic principle of counting, generalized basic principle of counting, permutation with repetition and without repetition, and combination with sequence consideration and without sequence consideration have been provided. The distribution of a set of elements into distinct groups and into indistinct groups has also been defined as a special application of the combination without sequence consideration. Some exercises throughout the chapter and problems at the end of the chapter have been provided to clarify the basic concepts. Some problems have also been presented to make connections to some well-known engineering problems such as *traveling salesperson (salesman) problem*, *0–1 knapsack problem*, and *job scheduling problem*.

This chapter will introduce the basic concepts of combinatorial analysis which is about counting the number of ways for sequencing or arranging the elements in a finite set or selecting a subset from a finite set based on some rules. While this topic is very important in its own sense with theoretical research areas and real-life implications; it is also fundamental for other topics in probability. As it will be especially clear in Chaps. 2 and 3, it is imperative to know the basics of combinatorial analysis to be able to perform even basic probability calculations. Thus, this chapter will provide the basic definitions such as the random experiment, the basic principle of counting, the generalized basic principle of counting, and permutation and combination that will be fundamental background for all chapters related to the basics of probability.

1.1 The Basic Principle of Counting and the Generalized Basic Principle of Counting

Example 1 Suppose that we roll a die. We certainly know that the possible outcomes are 1, 2, 3, 4, 5, 6. However, we don't know the real outcome with certainty before we roll the die.

© Springer Nature Switzerland AG 2019
E. Bas, *Basics of Probability and Stochastic Processes*,
https://doi.org/10.1007/978-3-030-32323-3_1

Definition 1 (*Random experiment*) A random experiment is an experiment for which the possible outcomes are certainly known, but the real outcome is *not known* with certainty before this experiment is performed.

Definition 2 (*Basic principle of counting*) Let n_1 be the number of the possible outcomes of random experiment 1, and let n_2 be the number of the possible outcomes of random experiment 2. Hence, the number of the possible outcomes as a result of performing random experiment 1 and random experiment 2 simultaneously will be

$$n_1 \, n_2$$

Example 1 (revisited 1) Note that rolling a die is a random experiment, and the number of the possible outcomes of rolling a die is 6. Suppose that we also toss a coin. We know that the possible outcomes for tossing a coin are Heads (H) and Tails (T). We consider Table 1.1 for the total number of the possible outcomes as a result of performing these two random experiments.

Definition 3 (*Generalized basic principle of counting*) Let n_1 be the number of the possible outcomes of random experiment 1, let n_2 be the number of the possible outcomes of random experiment 2, and generally let n_k be the number of the possible outcomes of random experiment k. Hence, the number of the possible outcomes as a result of performing these k random experiments simultaneously will be

$$n_1 n_2 \ldots n_k$$

Example 1 (revisited 2) Suppose that in addition to rolling a die and tossing a coin, we randomly select a ball from an urn that contains 1 white, 1 red, and 1 yellow ball. Accordingly, the number of the possible outcomes as a result of performing these three experiments will be

$$(6)(2)(3) = 36$$

Table 1.1 The number of the possible outcomes as a result of rolling a die and tossing a coin

Rolling a die	Tossing a coin	Rolling a die and tossing a coin
1	H, T	$(1, H), (1, T)$
2	H, T	$(2, H), (2, T)$
3	H, T	$(3, H), (3, T)$
4	H, T	$(4, H), (4, T)$
5	H, T	$(5, H), (5, T)$
6	H, T	$(6, H), (6, T)$
		$(6)(2) = 12$ possible outcomes

1.2 Combinatorial Analysis (Combinatorics)

In combinatorial analysis (combinatorics), we basically deal with the following two questions:

(a) In how many ways can a set of elements be *sequenced (arranged)*?
(b) In how many ways can a subset be *selected* from a set of elements?

Example 2 We consider four letters A, B, C, D.

Definition 4 (*Permutation without repetition*) Let n be the number of the *distinct* elements to be sequenced. The number of the possible sequences (arrangements) for these elements will be

$$n!$$

Example 2 (revisited 1) We consider Table 1.2 to find the number of the possible sequences of the letters A, B, C, D.

Remark 1 Any sequence (arrangement) is called as a *permutation*. As an example, (A, C, B, D) and (D, C, B, A) are two different permutations in Example 2 (revisited 1).

Definition 5 (*Permutation with repetition*) Let n be the number of the elements to be sequenced. Let also $n_i, i = 1, 2, \ldots, k$ be the number of the repetitions for each element i, where $n_1 + n_2 + \cdots + n_k = n$. Hence, the number of the possible sequences (arrangements) for these elements will be

$$\frac{n!}{n_1! n_2! \ldots n_k!}$$

Table 1.2 The number of the possible sequences of the letters A, B, C, D

1st position	2nd position	3rd position	4th position	Permutations
A	B	C	D	$(A, B, C, D), (A, B, D, C)$
		D	C	
	C	B	D	$(A, C, B, D), (A, C, D, B)$
		D	B	
	D	B	C	$(A, D, B, C), (A, D, C, B)$
		C	B	
B	$\ldots$	$\ldots$	$\ldots$	$\ldots$
C	$\ldots$	$\ldots$	$\ldots$	$\ldots$
D	$\ldots$	$\ldots$	$\ldots$	$\ldots$
4	3	2	1	$(4)(3)(2)(1) = 4! = 24$

Example   2 (revisited   2)   Suppose   that   we   have   ten   letters
$A, A, A, B, B, B, B, C, C, D$ to be sequenced. The number of the possible
sequences of these ten letters will be

$$\frac{10!}{3!4!2!1!} = 12,600$$

Definition 6 (*Combination with sequence consideration*) Let n be the number of
total elements, and let k be the number of the elements to be *selected* and *sequenced*.
The number of possible ways to *select* and *sequence* k elements out of n elements
will be

$$n(n-1)\ldots(n-(k-1)) = \frac{n(n-1)\ldots(n-(k-1))(n-k)!}{(n-k)!} = \frac{n!}{(n-k)!}$$

Definition 7 (*Combination without sequence consideration*) Let n be the number of
total elements, and let k be the number of the elements to be *selected*. The number
of possible ways to *select* k elements out of n elements will be

$$\frac{n(n-1)\ldots(n-(k-1))}{k!} = \frac{n(n-1)\ldots(n-(k-1))(n-k)!}{(n-k)!k!}$$

$$= \frac{n!}{(n-k)!k!} = \binom{n}{k}$$

Example 2 (revisited 3) Suppose that we wish to select 3 letters from four letters
A, B, C, D *with sequence consideration*. We consider Table 1.3 for the solution.

If we wish to select 3 letters from four letters A, B, C, D *without sequence consideration*, then the number of possible selections will be

Table 1.3 Possible ways to select 3 letters from the letters A, B, C, D with sequence consideration

1st position	2nd position	3rd position	Possible outcomes
A	B	C	$(A, B, C), (A, B, D)$
		D	
	C	B	$(A, C, B), (A, C, D)$
		D	
	D	B	$(A, D, B), (A, D, C)$
		C	
B	...	...	...
C	...	...	...
D	...	...	...
4	3	2	$(4)(3)(2) = \frac{4!}{(4-3)!} = 24$

$$\binom{4}{3} = \frac{4!}{(4-3)!3!} = 4$$

Definition 8 (*Distribution of n elements into k distinct groups*) Let n be the number of the elements, let k be the number of the *distinct* groups, and let n_i be the size of each *distinct* group i, $i = 1, 2, \ldots, k$, where $n_1 + n_2 + \cdots + n_k = n$. These n elements can be distributed to these k *distinct* groups in

$$\binom{n}{n_1}\binom{n-n_1}{n_2}\binom{n-n_1-n_2}{n_3} \cdots \binom{n-n_1-n_2-\cdots-n_{k-1}}{n_k}$$
$$= \frac{n!}{(n-n_1)!n_1!}\frac{(n-n_1)!}{(n-n_1-n_2)!n_2!} \cdots \frac{(n-n_1-n_2-\cdots-n_{k-1})!}{(n-n_1-n_2-\cdots-n_k)!n_k!}$$
$$= \frac{n!}{n_1!n_2! \ldots n_k!}$$

different ways.

Example 3 Suppose that there are 10 students, and three classes A, B, C. Class A has a capacity of 3 students, Class B has a capacity of 5 students, and Class C has a capacity of 2 students. We consider Table 1.4 to find the number of possible ways to distribute 10 students to these 3 classes.

Definition 9 (*Distribution of n elements into k indistinct groups*) Let n be the number of the elements, let k be the number of the *indistinct* groups, and let n_i be the size of each *indistinct* group i, $i = 1, 2, \ldots, k$, where $n_1 + n_2 + \cdots + n_k = n$. These n elements can be distributed to these k *indistinct* groups in

$$\frac{n!}{(n_1!n_2! \ldots n_k!)k!}$$

different ways.

Table 1.4 Distribution of 10 students to 3 classes

Class A (Capacity: 3)	Class B (Capacity: 5)	Class C (Capacity: 2)	Possible number of distributions
Total 10 students to be distributed to class A	Total $10 - 3 = 7$ students to be distributed to Class B	Total $7 - 5 = 2$ students to be distributed to Class C	
$\binom{10}{3}$	$\binom{7}{5}$	$\binom{2}{2}$	$\binom{10}{3}\binom{7}{5}\binom{2}{2}$
			$= \frac{10!}{3!5!2!} = 2{,}520$

Example 3 (revisited 1) Suppose now that 10 students are to be distributed to 3 indistinct classes with no titles. The number of possible ways to distribute 10 students to 3 classes will be

$$\frac{10!}{(3!5!2!)3!} = 420$$

Problems

1. 8 different customer orders arrive to a job shop to be machined by a turning lathe.
(a) How many different sequences are possible for machining if all 8 customer orders can be accepted?
(b) How many different sequences are possible for machining if only 4 customer orders can be accepted?

Solution

(a) This problem is a *permutation without repetition* problem as defined in Definition 4, thus

$$8! = 40,320$$

different sequences are possible for machining.
(b) This problem is a *combination with sequence consideration* problem as defined in Definition 6, thus

$$\frac{8!}{(8-4)!} = 1,680$$

different sequences are possible for machining.
2. There are 5 red balls, 4 yellow balls and 7 white balls in a box. In how many ways can these balls be arranged if the balls with the same color are supposed to be together?

Solution
There are

$$5! = 120$$

different arrangements of 5 red balls. There are

$$4! = 24$$

different arrangements of 4 yellow balls. There are

$$7! = 5,040$$

different arrangements of 7 white balls. Since there are 3 colors, there are

$$3! = 6$$

different arrangements of the colors. Finally, there are

$$3!(5!4!7!) = 6(120)(24)(5,040) = 87,091,200$$

different arrangements if the balls with the same color are supposed to be together.

3. A traveling salesman lives in Berlin (B), and wants to visit the customers in Hamburg (H), Essen (E), Munich (M), Dortmund (D), and Leipzig (L). He wants to visit each customer only once, and wants to return to his home city Berlin after he visits all customers. How many different *visiting sequences* are possible if the traveling salesman considers the total distance as the criterion?

Solution

The number of the visiting sequences is a *permutation without repetition* problem. Since there are 5 cities to be visited, there are

$$5! = 120$$

different visiting sequences. As an example, Fig. 1.1 shows the sequence $B - H - D - L - M - E - B$.

However, if the total distance is the criterion for the decision, then it is clear that the distance of each sequence will be the same with the distance of its reverse sequence, as an example $B - H - D - L - M - E - B$ and its reverse sequence $B - E - M - L - D - H - B$ yield the same total distances. As a result, there will be

$$\frac{5!}{2} = \frac{120}{2} = 60$$

different visiting sequences to be considered for the decision.

Remark 2 If the problem is to find the optimal visiting sequence which yields the *minimum total distance*, then the problem will be a very famous combinatorial optimization problem so-called *Traveling Salesperson (Salesman) Problem (TSP)*.

4. In a job shop, there are 5 jobs to be processed. Each job must first go to machine 1, then to machine 2. How many different processing sequences are possible?

Fig. 1.1 An example of
visiting sequences for
problem 3

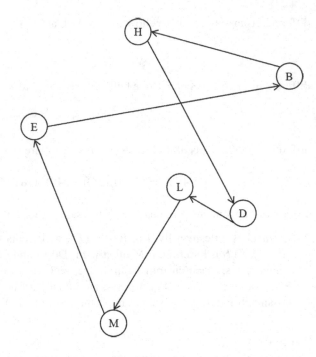

Solution

Since each job must first go to machine 1, then to machine 2 without changing the
sequence, there are

$$5! = 120$$

different processing sequences.

Remark 3 If the problem is to find the optimal processing sequence of jobs given a
specific criterion, then this problem will be a kind of *Job Scheduling Problem*.

5. There are 6 items to be considered to put into a knapsack. The knapsack has a
   capacity, each item has a weight, and a value. How many different groupings of
   items will be possible although all groupings of items may not be feasible due
   to the capacity constraint?

Solution

$$\sum_{i=1}^{6} \binom{6}{i} = 63$$

different groupings of items will be possible.

Remark 4 If the problem is to find the optimal grouping of items which yields the *maximum total value* while satisfying the *capacity constraint*, then it will be a very famous combinatorial optimization problem so-called 0–1 *Knapsack Problem*.

6. A student plans to buy 2 books from a collection of 6 Math (M), 7 Science (S), and 4 Economics (E) books. How many choices are possible if
(a) The books are to be on the same subject.
(b) The books are to be on different subjects.

Solution

(a) Since the books are to be on the same subject, the student can choose *either* 2 math books *or* 2 science books *or* 2 economics books.

There are

$$\binom{6}{2} = 15$$

different ways for the selection of *math* books. There are

$$\binom{7}{2} = 21$$

different ways for the selection of *science* books. There are

$$\binom{4}{2} = 6$$

different ways for the selection of *economics* books.

Since choosing math books *or* science books *or* economics books is *mutually exclusive*, there are

$$\binom{6}{2} + \binom{7}{2} + \binom{4}{2} = 15 + 21 + 6 = 42$$

different possible ways to select 2 books on the same subject.

(b) 2 subjects can be selected from a set of 3 subjects in

$$\binom{3}{2} = \frac{3!}{2!1!} = 3$$

different ways. These pairs of subjects are $M - S$, $M - E$, and $S - E$.

There are

$$\binom{6}{1}\binom{7}{1} = 42$$

different ways for the selection of 2 books from the pair of $M - S$ subjects. There are

$$\binom{6}{1}\binom{4}{1} = 24$$

different ways for the selection of 2 books from the pair of $M - E$ subjects. There are

$$\binom{7}{1}\binom{4}{1} = 28$$

different ways for the selection of 2 books from the pair of $S - E$ subjects. Since the selection of 2 books from each pair of subjects is *mutually exclusive*, there are

$$42 + 24 + 28 = 94$$

different ways to select 2 books on different subjects.

Remark 5 For the formal definition of the *mutually exclusive events*, please refer to Definition 4 in Chap. 2.

7. From a group of 8 red balls and 6 yellow balls, a group of 3 red balls and 3 yellow balls is to be built. Suppose that the balls are numbered. How many *different groups* are possible if yellow ball number 1 ($Y1$), and yellow ball number 4 ($Y4$) will *not* be in the same group?

Solution
There are three *mutually exclusive events*:

(i) $Y1$ will *be* in the group, but $Y4$ will *not be* in the group. Accordingly, there are

$$\binom{8}{3}\binom{4}{2} = 336$$

different selections.

(ii) $Y4$ will *be* in the group, but $Y1$ will *not be* in the group. Then there are

$$\binom{8}{3}\binom{4}{2} = 336$$

different selections.

(iii)   Both $Y1$ and $Y4$ will *not be* in the group. Hence, there are

$$\binom{8}{3}\binom{4}{3} = 224$$

different selections. As a result, there will be

$$336 + 336 + 224 = 896$$

different groups.

8.   There are 4 students to be paired. In how many ways can these students be paired if
(a)   Each pair will be defined as Pair A or Pair B?
(b)   Only making pairs is relevant without any pair titles?

Solution

(a)   Suppose that we name the students as S_1, S_2, S_3, S_4. We consider Table 1.5 for the solution, and find that 6 pairings are possible.

Alternatively, by referring to Definition 8, we can find that

$$\frac{4!}{2!2!} = 6$$

pairings are possible.

Table 1.5 Pairings *by* considering the pair titles

	Pair A	Pair B
1	S_1, S_2	S_3, S_4
2	S_1, S_3	S_2, S_4
3	S_1, S_4	S_2, S_3
4	S_2, S_3	S_1, S_4
5	S_2, S_4	S_1, S_3
6	S_3, S_4	S_1, S_2

Table 1.6 Pairings *without* considering the pair titles

1	S_1, S_2	S_3, S_4
2	S_1, S_3	S_2, S_4
3	S_1, S_4	S_2, S_3
3	S_2, S_3	S_1, S_4
2	S_2, S_4	S_1, S_3
1	S_3, S_4	S_1, S_2

(b) In this question, the pairs have no defined titles like Pair *A* or Pair *B*. Only the number of ways the students can be paired is relevant. We consider Table 1.6 for the solution, and find that 3 pairings are possible.

Alternatively, by referring to Definition 9, we can find that

$$\frac{4!}{(2!2!)2!} = 3$$

pairings are possible.

Chapter 2
Basic Concepts, Axioms and Operations in Probability

Abstract In this chapter, the definitions of the basic concepts in probability including the sample space, single event, event, mutually exclusive events, probability, and three axioms of probability have been provided. The basic operations in probability theory including the intersection, union, and complement have been given in comparison to the basic operations in set theory. The exercises throughout the chapter, and the problems at the end of the chapter elucidated the basic concepts in probability. In problems section, Venn diagrams have been used for the solutions, where relevant; and the solution of two well-known problems in probability, i.e. *birthday problem*, and *straight in a poker hand problem* was also presented.

This chapter will introduce the basic definitions, axioms, and the basic operations of probability. After learning the basic concepts of combinatorial analysis in Chap. 1; in this chapter, it will be possible to perform basic probability calculations, and the probability calculations based on the basic operations in probability theory. It will be clear that set theory has a strong relation with probability theory, and the basic operations in probability theory are the direct extensions of the basic operations in set theory. Thus, Venn diagrams will be used for illustration of the solutions of the examples and problems, where relevant. Without doubt, this chapter will provide some very important fundamentals such as the axioms of probability which can be implemented not only in other topics related to the basics of probability, but also in the topics related to the basics of stochastic processes.

2.1 Basic Concepts

Example 1 Suppose that we roll a die. We certainly know that the set of the possible outcomes is $S = \{1, 2, 3, 4, 5, 6\}$. At any roll, a possible outcome can be $\{1\}, \{2\}, \{3\}, \{4\}, \{5\}$ or $\{6\}$. We can define a subset of S as $A = \{2, 4, 6\}$ that includes the outcomes with even values. We can also define another subset of S as $B = \{1, 3, 5\}$ that includes the outcomes with odd values.

© Springer Nature Switzerland AG 2019
E. Bas, *Basics of Probability and Stochastic Processes*,
https://doi.org/10.1007/978-3-030-32323-3_2

Definition 1 (*Sample space*) *Sample space* is the set of *all possible* outcomes of a random experiment.

Definition 2 (*Single event*) *Single event* is each possible *single* outcome of a random experiment.

Definition 3 (*Event*) *Event* is a *subset* of a sample space of a random experiment.

Definition 4 (*Mutually exclusive events*) Two or more events are said to be *mutually exclusive* if their intersection is a *null set*.

Remark 1 Each *single event* can also be defined as an *event*, and each *event* can consist of one single event or more single events.

Example 1 (revisited 1) For the random experiment of rolling a die, $S = \{1, 2, 3, 4, 5, 6\}$ is the sample space. $\{1\}, \{2\}, \{3\}, \{4\}, \{5\}$ and $\{6\}$ are the single events. A and B are mutually exclusive events since $A \cap B = \emptyset$.

Definition 5 (*Probability*) Let n be the number of trials of a random experiment, and $n(A)$ be the number of trials that result in event A out of n trials. Hence, the probability of event A can be defined as

$$P(A) = \frac{n(A)}{n} \text{ as } n \to \infty$$

WARNING Definition 5 is one of the probability definitions which is based on *relative frequency*. There are also more rigorous definitions of probability.

Example 1 (revisited 2) We assume that it is a *fair die*, i.e. each single event is *equally likely* to appear. Accordingly, the probability of each single event will be

$$P(\{1\}) = P(\{2\}) = P(\{3\}) = P(\{4\}) = P(\{5\}) = P(\{6\}) = \frac{1}{6}$$

2.2 Axioms of Probability

Axiom 1 For any event A of a random experiment,

$$0 \leq P(A) \leq 1$$

holds.

Axiom 2 For each random experiment,

$$P(S) = 1$$

holds.

Axiom 3 The probability of the union of the mutually exclusive events will be

$$P\left(\bigcup_{i=1}^{n} A_i\right) = \sum_{i=1}^{n} P(A_i)$$

where A_i, $i = 1, 2, \ldots, n$ are mutually exclusive events, i.e. $\bigcap_{i=1}^{n} A_i = \emptyset$.

Example 1 (revisited 3) Since there has to be an outcome at any roll,

$$P(\{1\} \cup \{2\} \cup \{3\} \cup \{4\} \cup \{5\} \cup \{6\}) = P(S) = 1$$

by Axiom 2. The probability of rolling an odd number will be

$$P(\{1\} \cup \{3\} \cup \{5\}) = P(\{1\}) + P(\{3\}) + P(\{5\}) = \frac{1}{6} + \frac{1}{6} + \frac{1}{6} = \frac{3}{6} = \frac{1}{2}$$

by Axiom 3, since the single events $\{1\}$, $\{3\}$, $\{5\}$ are *mutually exclusive* events.

2.3 Basic Operations in Set Theory Versus Basic Operations in Probability Theory

The basic operations in set theory, and the basic operations in probability theory are provided in comparison to each other in Table 2.1.

Example 1 (revisited 4) The probability that the outcome is *not* a *prime* number will be

$$P\left(D^C\right) = 1 - P(D)$$

where

D: Event that the outcome is a *prime* number

Since $D = \{2, 3, 5\}$,

$$P(D) = \frac{3}{6} = \frac{1}{2}$$

Finally,

$$P\left(D^C\right) = 1 - P(D) = 1 - \frac{1}{2} = \frac{1}{2}$$

Example 1 (revisited 5) The probability that the outcome is an *even* number *or* a *prime* number will be

Table 2.1 Basic operations in set theory versus basic operations in probability theory

Basic operations in set theory	Basic operations in probability theory
Let	Let
A,B: Two sets	*A,B:* Two events of a random experiment
U : Universal set	S : Sample space
$n(.)$: The number of elements in a set	$P(.)$: Probability of an event
$A \cap B$: Intersection of the sets A and B	$A \cap B$: Intersection of the events A and B
$A \cup B$: Union of the sets A and B	$A \cup B$: Union of the events A and B
A^C, B^C: Complement of the set A, complement of the set B	A^C, B^C: Complement of the event A, complement of the event B
Hence, the following equalities and inequalities will hold:	Hence, the following equalities and inequalities will hold:
$n(A \cup B) = n(A) + n(B) - n(A \cap B)$	$P(A \cup B) = P(A) + P(B) - P(A \cap B)$
$n\left(A^C\right) = n(U) - n(A)$	$P\left(A^C\right) = P(S) - P(A) = 1 - P(A)$
$n\left(B^C\right) = n(U) - n(B)$	$P\left(B^C\right) = P(S) - P(B) = 1 - P(B)$
If $A \subseteq B$	If $A \subseteq B$
$n(A) \leq n(B)$	$P(A) \leq P(B)$
$n(A \cup B \cup C) = n(A) + n(B) + n(C)$ $- n(A \cap B) - n(A \cap C)$ $- n(B \cap C) + n(A \cap B \cap C)$	$P(A \cup B \cup C) = P(A) + P(B) + P(C)$ $- P(A \cap B) - P(A \cap C)$ $- P(B \cap C) + P(A \cap B \cap C)$

$$P(A \cup D) = P(A) + P(D) - P(A \cap D)$$

where

A: Event that the outcome is an *even* number
D: Event that the outcome is a *prime* number

Since $A = \{2, 4, 6\}$ and $D = \{2, 3, 5\}$, it follows that $(A \cap D) = \{2\}$, and

$$P(A) = \frac{3}{6} = \frac{1}{2}$$
$$P(D) = \frac{3}{6} = \frac{1}{2}$$
$$P(A \cap D) = \frac{1}{6}$$

Finally, the probability that the outcome is an *even* number *or* a *prime* number will be

$$P(A \cup D) = P(A) + P(D) - P(A \cap D) = \frac{1}{2} + \frac{1}{2} - \frac{1}{6} = \frac{5}{6}$$

Remark 2 Note that the union of the events A and D means that *at least A or D* occurs, i.e. (1) only A occurs, *or* (2) only D occurs, *or* (3) both A and D occur.

Example 2 Consider two *mutually exclusive* events A and B with probabilities $P(A) = 0.4$ and $P(B) = 0.5$. The probability that both A *and* B occur will be

$$P(A \cap B) = 0$$

since they are *mutually exclusive* events, and the probability that *at least A or B* occurs will be

$$P(A \cup B) = P(A) + P(B) = 0.4 + 0.5 = 0.9$$

by Axiom 3.

Problems

1.   A coin is tossed, and a die is rolled. Let A be the event that the outcome of tossing a coin will be Heads (H), and the outcome of rolling a die will be an odd number. Determine the sample space, and the event A.

Solution

$$S = \{(H, 1), (H, 2), (H, 3), (H, 4), (H, 5), (H, 6),$$
$$(T, 1), (T, 2), (T, 3), (T, 4), (T, 5), (T, 6)\}$$
$$A = \{(H, 1), (H, 3), (H, 5)\}$$

2.   There are three components of a machine. The following variable is defined for each component $i = 1, 2, 3$:

$$I_i = \begin{cases} 1 \text{ if component } i \text{ is on} \\ 0 \text{ if component } i \text{ is off} \end{cases}$$

Determine the sample space, and the event A that *at least two* components are on.

Solution

$$S = \{(1, 1, 1), (1, 1, 0), (1, 0, 1), (0, 1, 1), (1, 0, 0), (0, 1, 0), (0, 0, 1), (0, 0, 0)\}$$
$$A = \{(1, 1, 1), (1, 1, 0), (1, 0, 1), (0, 1, 1)\}$$

Remark 3 Note that $n(S)$ could also be calculated as

$$\binom{3}{0} + \binom{3}{1} + \binom{3}{2} + \binom{3}{3} = 1 + 3 + 3 + 1 = 8$$

or as

$$2^3 = 8$$

by using the *generalized principle of counting.* $n(A)$ could also be calculated as

$$\binom{3}{2} + \binom{3}{3} = 3 + 1 = 4$$

3. In a class of 50 students, each student speaks at least one of the languages English, French, German. 5 students speak all three languages, while 15 students speak both French and German, 17 students speak both French and English, 13 students speak both German and English. In this class, there are also 3 students who can only speak French, 2 students who can only speak German, and 10 students who can only speak English. If a student is selected randomly from this class, what is the probability that this student speaks *at least* English *or* German?

Solution

Let

E: Event that a randomly selected student speaks *English*
F: Event that a randomly selected student speaks *French*
G: Event that a randomly selected student speaks *German*

Hence, the probability that a randomly selected student speaks *at least* English *or* German will be

$$P(E \cup G) = P(E) + P(G) - P(E \cap G)$$

where

$$P(E) = \frac{35}{50}$$
$$P(G) = \frac{25}{50}$$
$$P(E \cap G) = \frac{13}{50}$$

by considering the Venn diagram in Fig. 2.1.

Fig. 2.1 Venn diagram for problem 3

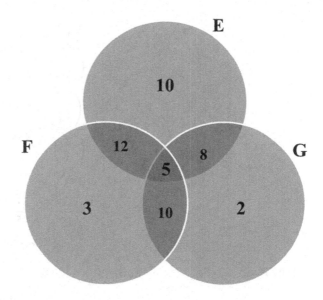

Finally,

$$P(E \cup G) = P(E) + P(G) - P(E \cap G) = \frac{35}{50} + \frac{25}{50} - \frac{13}{50} = \frac{47}{50}$$

4. There are 5 red balls, 7 yellow balls, and 3 white balls in an urn, and 3 balls are randomly withdrawn from this urn. Find the probability that

(a) All withdrawn balls are of the *same color*.
(b) All withdrawn balls are of *different color*.
(c) At least 2 yellow balls are withdrawn.
(d) 1 red ball *or* 2 white balls are withdrawn.

Solution

Let

R: Number of the *red* balls withdrawn
Y: Number of the *yellow* balls withdrawn
W: Number of the *white* balls withdrawn

(a) The probability that all withdrawn balls are of the *same color* will be

$$P(\{R = 3\} \cup \{Y = 3\} \cup \{W = 3\}) = P(\{R = 3\}) + P(\{Y = 3\}) + P(\{W = 3\})$$

by Axiom 3, where

$$P(\{R = 3\}) = \frac{\binom{5}{3}}{\binom{15}{3}} = \frac{10}{455}$$

$$P(\{Y = 3\}) = \frac{\binom{7}{3}}{\binom{15}{3}} = \frac{35}{455}$$

$$P(\{W = 3\}) = \frac{\binom{3}{3}}{\binom{15}{3}} = \frac{1}{455}$$

Finally,

$$P(\{R = 3\} \cup \{Y = 3\} \cup \{W = 3\}) = \frac{10}{455} + \frac{35}{455} + \frac{1}{455} = \frac{46}{455}$$

(b) The probability that all withdrawn balls are of *different color* will be

$$P(\{R = 1\} \cap \{Y = 1\} \cap \{W = 1\}) = \frac{\binom{5}{1}\binom{7}{1}\binom{3}{1}}{\binom{15}{3}} = \frac{105}{455}$$

(c) The probability that at least 2 yellow balls are withdrawn will be

$$P(Y \geq 2) = P(Y = 2) + P(Y = 3)$$

where

$$P(Y = 2) = \frac{\binom{7}{2}\binom{5}{1}}{\binom{15}{3}} + \frac{\binom{7}{2}\binom{3}{1}}{\binom{15}{3}} = \frac{105}{455} + \frac{63}{455} = \frac{168}{455}$$

$$P(Y = 3) = \frac{\binom{7}{3}}{\binom{15}{3}} = \frac{35}{455}$$

Finally,

$$P(Y \geq 2) = P(Y = 2) + P(Y = 3) = \frac{168}{455} + \frac{35}{455} = \frac{203}{455}$$

(d) The probability that 1 red ball *or* 2 white balls are withdrawn will be

$$P(\{R = 1\} \cup \{W = 2\}) = P(\{R = 1\}) + P(\{W = 2\}) - P(\{R = 1\} \cap \{W = 2\})$$

where

$$P(R = 1) = \frac{\binom{5}{1}\binom{7}{2}}{\binom{15}{3}} + \frac{\binom{5}{1}\binom{3}{2}}{\binom{15}{3}} + \frac{\binom{5}{1}\binom{7}{1}\binom{3}{1}}{\binom{15}{3}} = \frac{225}{455}$$

$$P(W = 2) = \frac{\binom{3}{2}\binom{5}{1}}{\binom{15}{3}} + \frac{\binom{3}{2}\binom{7}{1}}{\binom{15}{3}} = \frac{36}{455}$$

$$P(\{R = 1\} \cap \{W = 2\}) = \frac{\binom{5}{1}\binom{3}{2}}{\binom{15}{3}} = \frac{15}{455}$$

Finally,

$$P(\{R = 1\} \cup \{W = 2\}) = \frac{225}{455} + \frac{36}{455} - \frac{15}{455} = \frac{246}{455}$$

Remark 4 The basic difference in the solutions of (a) and (b) is that in (a), there are three *mutually exclusive events*, while in (b) the *generalized basic principle of counting* is used in the numerator of probability for the number of the possible selections of 1 red, 1 yellow *and* 1 white ball. In the solution of (c), for finding $P(Y = 2)$, two *mutually exclusive events* are considered, i.e. selecting the other ball as a red ball *or* as a white ball. The explanation of (d) will be analogous.

5. There are 25 students in a classroom. What is the probability that *each student* has a *different birthday*? What is the probability that *three students* have the same birthday?

Solution

The probability that *each student* has a *different birthday* will be

$$\frac{(365)(364)(363)\ldots(365-(25-1))}{365^{25}} = \frac{(365)(364)(363)\ldots(341)}{365^{25}}$$

$$= \frac{\frac{(365)(364)(363)\ldots(341)(340!)}{340!}}{365^{25}} = \frac{\left(\frac{365!}{340!}\right)}{365^{25}} = 0.4313$$

The probability that *three students* have the same birthday will be

$$\frac{\binom{25}{3}(365)(364)(363)(362)\ldots(364-(22-1))}{365^{25}}$$

$$= \frac{\binom{25}{3}(365)(364)(363)(362)\ldots(343)}{365^{25}}$$

$$= \frac{\binom{25}{3}\frac{(365)(364)(363)(362)\ldots(343)(342!)}{342!}}{365^{25}}$$

$$= \frac{\binom{25}{3}\left(\frac{365!}{342!}\right)}{365^{25}} = 0.0085$$

Remark 5 3 students who have the same birthday can be selected by using *combination without sequence consideration*. These three students can have the same birthday from any of 365 days. Each of the remaining 22 students can have one of the remaining 364 birthdays given that they have different birthdays.

6. What is the probability of a *straight* in a *poker* hand? (*Note*: A poker hand consists of 5 cards. If the cards are in ascending order of rank, but not all of them have the same suit, that poker hand is called as a *straight*.)

Solution

The probability of a straight in a poker hand will be

$$\frac{10\left(4^5-4\right)}{\binom{52}{5}} = 0.0039$$

Remark 6 Note that there are total 52 cards in a deck, and 4 suits called as diamonds, clubs, hearts, and spades for each card. In a poker hand of 5 cards, 10 different ways will be possible with an ascending order of rank such that

$$A - 2 - 3 - 4 - 5$$
$$2 - 3 - 4 - 5 - 6$$
$$3 - 4 - 5 - 6 - 7$$
$$\cdots\cdots$$
$$10 - J - Q - K - A$$

Note also that $(4^5 - 4)$ implies the number of the possible outcomes regarding suits for each ascending order of rank, in which *not* all of the cards have the same suit.

7. Let A, B, C be three mutually exclusive events with respective probabilities $\frac{1}{3}, \frac{1}{3}, \frac{1}{4}$.

   (a) What is the probability that either A or B will occur, but C will *not* occur?
   (b) What is the probability that *only* B will occur?
   (c) What is the probability that none of A, B *or* C will occur?

Solution
We consider the Venn diagrams in Fig. 2.2 for the solution.

(a) The probability that either A or B will occur, but C will *not* occur will be

$$P\big(\{A \cup B\} \cap C^C\big) = P(A \cup B) = P(A) + P(B) = \frac{1}{3} + \frac{1}{3} = \frac{2}{3}$$

by Axiom 3.

(b) The probability that only B will occur will be

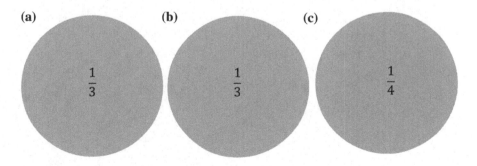

(a)   **(b)**   **(c)**

$\frac{1}{3}$ $\frac{1}{3}$ $\frac{1}{4}$

Fig. 2.2 Venn diagrams for problem 7

$$P(B) = \frac{1}{3}$$

(c) The probability that none of A, B or C will occur will be

$$1 - P(A \cup B \cup C) = 1 - (P(A) + P(B) + P(C))$$
$$= 1 - \left(\frac{1}{3} + \frac{1}{3} + \frac{1}{4}\right) = \frac{1}{12}$$

Chapter 3
Conditional Probability, Bayes' Formula, Independent Events

Abstract In this chapter, three important topics, i.e. the conditional probability, Bayes's formula, and the independence of two and three events have been introduced, which are crucial not only for the basics of probability, but also for the basics of stochastic processes. Some illustrative examples and problems have been presented, and most of the solutions have been supported by using the tree diagrams for a better comprehension.

This chapter will provide three very important topics in probability and stochastic processes, i.e. conditional probability, Bayes' formula, and independent events. After learning the basic concepts, axioms, and the operations in probability in Chap. 2; in this chapter, it will be possible to update the probability calculations given the occurrence of another event by using the *conditional probability* formula and *Bayes's formula*. The tree diagrams will be used for a better representation, and comprehension of the examples and problems related to the conditional probability. One extremely important assumption in probability and stochastic processes is the *independence of events*. Thus, the formulae for the independence of two and three events will be introduced, and the importance of this assumption will be illustrated with some examples and problems.

3.1 Conditional Probability

Example 1 Suppose that we roll two dice. *Given that* the sum of the outcomes of two dice is an even number, what is the probability that the outcome of the first die is an odd number?

There are two events in this example that can be defined as follows:

A: Event that the outcome of the first die is an odd number
B: Event that the sum of the outcomes of two dice is an even number

Hence, the question will be:

$$P(A|B) =? \text{ rather than } P(A) =?$$

© Springer Nature Switzerland AG 2019
E. Bas, *Basics of Probability and Stochastic Processes*,
https://doi.org/10.1007/978-3-030-32323-3_3

Definition 1 (*Conditional probability*)

$$P(A|B) = \frac{P(A \cap B)}{P(B)}$$

is called as the *conditional probability* formula, where
   $P(A|B)$: Probability of event A *given that* event B occurs

Example 1 (revisited 1) We can define the sample space S, and the events A, B for this problem as follows:

$$S = \{(1, 1), (1, 2), (1, 3), \ldots, (6, 5), (6, 6)\}$$
$$n(S) = 36$$

$$A = \{(1, 1), (1, 2), (1, 3), (1, 4), (1, 5), (1, 6), (3, 1), (3, 2),$$
$$(3, 3), (3, 4), (3, 5), (3, 6), (5, 1), (5, 2), (5, 3), (5, 4), (5, 5), (5, 6)\}$$
$$n(A) = 18$$

$$B = \{(1, 1), (1, 3), (1, 5), (2, 2), (2, 4), (2, 6), (3, 1), (3, 3), (3, 5),$$
$$(4, 2), (4, 4), (4, 6), (5, 1), (5, 3), (5, 5), (6, 2), (6, 4), (6, 6)\}$$
$$n(B) = 18$$

Accordingly,

$$A \cap B = \{(1, 1), (1, 3), (1, 5), (3, 1), (3, 3), (3, 5), (5, 1), (5, 3), (5, 5)\}$$
$$n(A \cap B) = 9$$

As a result, the probabilities $P(A \cap B)$, $P(B)$, and finally $P(A|B)$ can be calculated as follows:

$$P(A \cap B) = \frac{n(A \cap B)}{n(S)} = \frac{9}{36} = \frac{1}{4}$$

$$P(B) = \frac{n(B)}{n(S)} = \frac{18}{36} = \frac{1}{2}$$

$$P(A|B) = \frac{P(A \cap B)}{P(B)} = \frac{\left(\frac{1}{4}\right)}{\left(\frac{1}{2}\right)} = \frac{1}{2}$$

Remark 1 $n(S)$ could easily be calculated by using *the basic principle of counting* as

$$n(S) = (6)(6) = 36$$

Definition 2 (*Multiplication rule*)

$$P(E_1 \cap E_2 \cap E_3 \cap \ldots \cap E_n)$$
$$= P(E_1)P(E_2|E_1)P(E_3|E_1 \cap E_2) \ldots P(E_n|E_1 \cap E_2 \cap \ldots \cap E_{n-1})$$

is called as the *multiplication rule*, where

$$E_1, E_2, E_3, \ldots, E_n$$

are events.

Example 2 There are 4 red balls, 5 green balls and 2 black balls in a urn. A ball is withdrawn from this urn three times *without replacement*. What is the probability that the first two balls withdrawn are red, and the third ball withdrawn is green?
   Let

R_i: Event that a *red* ball is obtained at ith withdrawal
G_i: Event that a *green* ball is obtained at ith withdrawal
B_i: Event that a *black* ball is obtained at ith withdrawal

   The tree diagram in Fig. 3.1 illustrates this random experiment with probability values on the branches.

   By following the branches of the *tree diagram*, the probability that the first two balls withdrawn are red, and the third ball withdrawn is green will be

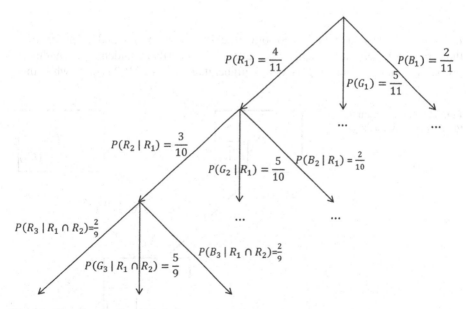

Fig. 3.1 Tree diagram for example 2

$$P(R_1 \cap R_2 \cap G_3) = \left(\frac{4}{11}\right)\left(\frac{3}{10}\right)\left(\frac{5}{9}\right) = \frac{2}{33}$$

By using the *multiplication rule formula*, the result will also be

$$P(R_1 \cap R_2 \cap G_3) = P(R_1)P(R_2|R_1)P(G_3|R_1 \cap R_2)$$

$$= \left(\frac{4}{11}\right)\left(\frac{3}{10}\right)\left(\frac{5}{9}\right) = \frac{2}{33}$$

3.2 Bayes' Formula

Definition 3 (*Total probability formula*) Let A and B be two events. Hence, the total probability formula can be defined as

$$P(A) = P(A|B)P(B) + P\left(A|B^C\right)P\left(B^C\right)$$

and can be illustrated as in Fig. 3.2.

Given that $B_1, B_2, \ldots, B_n$ are *mutually exclusive* events, and $S = B_1 \cup B_2 \cup \cdots \cup B_n$, *total probability formula* can also be generalized as

$$P(A) = \sum_{i=1}^{n} P(A|B_i)P(B_i)$$

Example 3 Each student in a classroom either loves math or not, and the probability that a randomly selected student loves math is 0.7. Given that a student loves math, the probability that this student gets a mark higher than 80 out of 100 is 0.85, while this

Fig. 3.2 Tree diagram for total probability formula

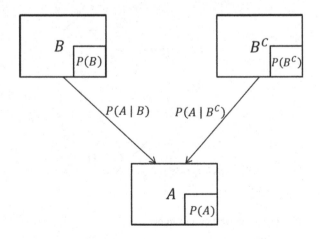

Fig. 3.3 Tree diagram for example 3

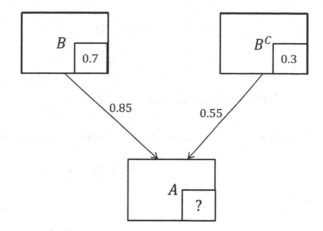

probability is 0.55 given that this student doesn't love math. What is the probability that a randomly selected student gets a mark higher than 80 out of 100?

Let

B: Event that a randomly selected student loves math
A: Event that a randomly selected student gets a mark higher than 80 out of 100

Accordingly, the problem can be illustrated as given in Fig. 3.3.

$$P(A) = P(A|B)P(B) + P(A|B^C)P(B^C)$$
$$= (0.85)(0.7) + (0.55)(0.3) = 0.76$$

Definition 4 (*Bayes' formula*) Let A and B be two events. Hence, Bayes' formula can be defined as

$$P(B|A) = \frac{P(B \cap A)}{P(A)} = \frac{P(A|B)P(B)}{P(A|B)P(B) + P(A|B^C)P(B^C)}$$

Given that $B_1, B_2, \ldots, B_n$ are *mutually exclusive* events, and $S = B_1 \cup B_2 \cup \cdots \cup B_n$, *Bayes' formula* can also be generalized as

$$P(B_i|A) = \frac{P(A|B_i)P(B_i)}{\sum_{i=1}^{n} P(A|B_i)P(B_i)} \quad i = 1, 2, \ldots, n$$

Example 3 (revisited 1) Given that a randomly selected student gets a mark higher than 80 out of 100, what is the probability that this student loves math?

$$P(B|A) = \frac{P(A|B)P(B)}{P(A|B)P(B) + P(A|B^C)P(B^C)}$$

$$= \frac{(0.85)(0.7)}{(0.85)(0.7) + (0.55)(0.3)} = 0.7829$$

3.3 Independent Events

Definition 5 (*Independence of two events*) A and B are two *independent* events, if

$$P(A|B) = P(A)$$

and

$$P(B|A) = P(B)$$

hold, which mean

$$\frac{P(A \cap B)}{P(B)} = P(A)$$

$$\frac{P(B \cap A)}{P(A)} = P(B)$$

Finally, if

$$P(A \cap B) = P(A)P(B)$$

A and B are *independent* events.

Definition 6 (*Independence of three events*) A, B and C are three *independent* events, if

$$P(A \cap B \cap C) = P(A)P(B)P(C)$$
$$P(A \cap B) = P(A)P(B)$$
$$P(A \cap C) = P(A)P(C)$$
$$P(B \cap C) = P(B)P(C)$$

Remark 2 We can explain the *independence of events* also by intuition. Independence has the general meaning "*having no effect*". If two or more events are *independent*, then having knowledge about the occurrence of one event should have *no effect* on the probability of the occurrence of the other event(s). If two or more events are *not independent*, then they are said to be *dependent*.

Example 1 (revisited 2) Are the events A and B *independent*?
We can calculate $P(A)$ by using Example 1 (revisited 1) as follows:

$$P(A) = \frac{n(A)}{n(S)} = \frac{18}{36} = \frac{1}{2}$$

Recall that we have already calculated $P(B)$ and $P(A \cap B)$ in Example 1 (revisited 1) as follows:

$$P(B) = \frac{n(B)}{n(S)} = \frac{18}{36} = \frac{1}{2}$$

$$P(A \cap B) = \frac{n(A \cap B)}{n(S)} = \frac{9}{36} = \frac{1}{4}$$

Since

$$P(A \cap B) = \frac{1}{4} = P(A)P(B) = \left(\frac{1}{2}\right)\left(\frac{1}{2}\right) = \frac{1}{4}$$

we conclude that A and B are *independent* events.

Example 4 We consider a *parallel system* with 3 components. If each component functions with probabilities $0.1, \ 0.5, 0.4$ *independently* from each other, what is the probability that the system functions? (*Note*: A parallel system functions if *at least one* of its components functions.)

Let

A: Event that the system functions
A^C: Event that the system does not function
C_i: Event that component i functions, $i = 1, 2, 3$
C_i^C: Event that component i does not function, $i = 1, 2, 3$

Note that

$$P(A) = 1 - P\left(A^C\right)$$

For a parallel system with 3 components,

$$P\left(A^C\right) = P\left(C_1^C \cap C_2^C \cap C_3^C\right)$$

since a parallel system does not function if no component functions. Since components function *independently* from each other,

$$P\left(A^C\right) = P\left(C_1^C \cap C_2^C \cap C_3^C\right) = P\left(C_1^C\right)P\left(C_2^C\right)P\left(C_3^C\right)$$
$$= (1 - P(C_1))(1 - P(C_2))(1 - P(C_3))$$

$$= (1 - 0.1)(1 - 0.5)(1 - 0.4) = (0.9)(0.5)(0.6) = 0.27$$

Finally, the probability that the system functions will be

$$P(A) = 1 - P(A^C) = 1 - 0.27 = 0.73$$

Problems

1. A coin with heads probability $\frac{1}{3}$ is tossed for 10 times *independently* from each other.
(a) What is the probability that no heads will appear?
(b) Given that the first toss results in heads, what is the probability that the second toss results also in heads?

Solution
Let

H_i: Event that the outcome of the ith toss will be heads

(a) The probability that no heads will appear will be

$$P\left(H_1^C \cap H_2^C \ldots H_{10}^C\right) = P\left(H_1^C\right)P\left(H_2^C\right)\ldots P\left(H_{10}^C\right)$$
$$= (1 - P(H_1))(1 - P(H_2))\ldots(1 - P(H_{10}))$$
$$= \left(1 - \frac{1}{3}\right)\left(1 - \frac{1}{3}\right)\ldots\left(1 - \frac{1}{3}\right) = \left(\frac{2}{3}\right)^{10} = 0.0173$$

by *independence* of outcomes.

(b) The probability that the second toss results also in heads given that the first toss results in heads will be

$$P(H_2|H_1) = P(H_2) = \frac{1}{3}$$

by *independence* of outcomes.

2. Please recall Problem 3 from Chap. 2. Given that a randomly selected student can speak English, what is the probability that this student can also speak German?

Solution
By using the Venn diagram representation in Fig. 2.1,

$$P(G|E) = \frac{P(G \cap E)}{P(E)} = \frac{\left(\frac{13}{50}\right)}{\left(\frac{35}{50}\right)} = \frac{13}{35}$$

3. Please recall from Problem 4 in Chap. 2 that there are 5 red balls, 7 yellow balls, and 3 white balls in an urn, and 3 balls are randomly selected from this urn. Given that 1 selected ball is white, what is the probability that 2 other balls are red?

Solution

$$P(\{R = 2\}|\{W = 1\}) = \frac{P(\{R = 2\} \cap \{W = 1\})}{P(W = 1)}$$

$$= \frac{\dfrac{\binom{5}{2}\binom{3}{1}}{\binom{15}{3}}}{\dfrac{\binom{3}{1}\binom{5}{2}}{\binom{15}{3}} + \dfrac{\binom{3}{1}\binom{7}{2}}{\binom{15}{3}} + \dfrac{\binom{3}{1}\binom{5}{1}\binom{7}{1}}{\binom{15}{3}}}$$

$$= \frac{30}{198} = \frac{5}{33}$$

4. In a university with a total of 15,000 students, 1,500 students study Industrial Engineering (IE). 12% of IE students are left-handed, while 15% of students from other majors are left-handed. If a student is randomly selected and found to be right-handed, what is the probability that this student studies a major other than IE?

Solution
Let

E: Event that a randomly selected student studies IE
E^C: Event that a randomly selected student studies a major other than IE
R: Event that a randomly selected student is right-handed

By noting that $P(E) = \frac{1,500}{15,000} = 0.1$, we can construct the tree diagram in Fig. 3.4 for the solution.

Accordingly, the probability that a randomly selected student studies a major other than IE given that this student is right-handed will be

$$P(E^C|R) = \frac{P(R|E^C)P(E^C)}{P(R|E)P(E) + P(R|E^C)P(E^C)}$$

$$= \frac{(0.85)(0.9)}{(0.88)(0.1) + (0.85)(0.9)} = 0.8968$$

Fig. 3.4 Tree diagram for problem 4

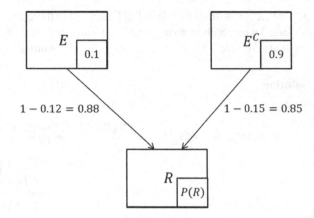

5. If it is rainy, the probability that there will be at least one accident on a given road is 0.55, while this probability is 0.15 for non-rainy days. The probability that it will be rainy tomorrow is 0.6. What is the probability that there will be *no accident* on that given road tomorrow? Given that there is at least one accident on that given road, what is the probability that it will *not* rain tomorrow?

Solution

Let

R: Event that it will be rainy tomorrow
R^C: Event that it will not be rainy tomorrow
A: Event that there will be at least one accident on that given road tomorrow

We can construct the tree diagram in Fig. 3.5 for the solution.

The probability that there will be *at least one accident* on that given road tomorrow will be

Fig. 3.5 Tree diagram for problem 5

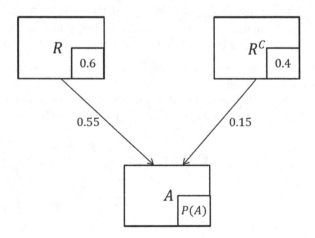

$$P(A) = P(A|R)P(R) + P(A|R^C)P(R^C)$$
$$= (0.55)(0.6) + (0.15)(0.4) = 0.39$$

Accordingly, the probability that there will be *no accident* on that given road tomorrow will be

$$P(A^C) = 1 - P(A) = 1 - 0.39 = 0.61$$

Given that there is *at least one accident* on that given road, the probability that it will *not* rain tomorrow will be

$$P(R^C|A) = \frac{P(A|R^C)P(R^C)}{P(A)} = \frac{(0.15)(0.4)}{0.39} = 0.1538$$

6. The stock price of a company is equally likely to have two possible values $5 or $8 in the initial day. Each other day, the stock price is assumed to increase by 10% with probability $\frac{1}{3}$ or decrease by 5% with probability $\frac{2}{3}$ with respect to the previous day. What is the probability that the sum of the stock prices will exceed $25 in the first three days?

Solution
We can construct the tree diagram in Fig. 3.6 for the solution of the problem.

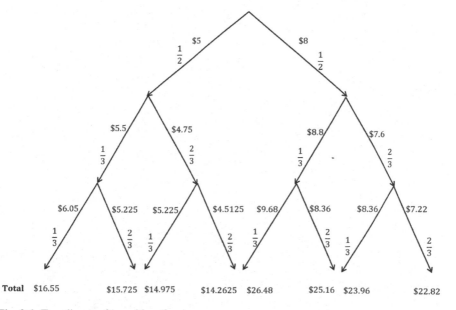

Fig. 3.6 Tree diagram for problem 6

There are two possible values that exceed \$25, i.e. \$26.48 and \$25.16. By following the branches of the tree diagram leading to these values, the probability that the sum of the stock prices will exceed \$25 in the first three days will be

$$\left(\frac{1}{2}\right)\left(\frac{1}{3}\right)\left(\frac{1}{3}\right) + \left(\frac{1}{2}\right)\left(\frac{1}{3}\right)\left(\frac{2}{3}\right) = \frac{1}{6}$$

Chapter 4
Introduction to Random Variables

Abstract In this chapter, the basic concepts for both discrete and continuous random variables were introduced. The definition of a random variable, a discrete random variable, and a continuous random variable, and the formulae of the probability mass function, probability density function, cumulative distribution function, complementary cumulative distribution function (tail function), expected value of a random variable, expected value of a function of a random variable, and the variance and standard deviation of a random variable have been provided. In addition to the solutions of some typical examples and problems for the discrete and continuous random variables, one well-known problem, i.e. *newsboy* (*newsvendor*) *problem* has also been presented.

This chapter will introduce one of the major areas in probability and stochastic processes, i.e. random variables. It is clear from Chaps. 1 and 2 that the enumeration of the outcomes or finding the possible number of outcomes can be crucial. However, sometimes not the outcomes, but a real-valued function defined on the sample space can be significant. Thus, we define random variables that can be *discrete* or *continuous*, which can be determined based on the countability of the number of the possible values. In this chapter, the basic parameters will be provided that are relevant for any discrete and continuous random variables, which will prove to be extremely important not only for the remaining chapters related to the basics of probability, but also for the chapters related to the basics of stochastic processes. Several examples and problems will be provided to draw attention to this major area.

4.1 Random Variables

Example 1 Suppose that we toss two coins. However, we are not interested in the outcomes, rather in the total number of heads. Then we can define the function illustrated in Fig. 4.1.

Definition 1 (*Random variable*) A *real-valued function* defined on the *sample space* of a random experiment is called as a *random variable*.

© Springer Nature Switzerland AG 2019
E. Bas, *Basics of Probability and Stochastic Processes*,
https://doi.org/10.1007/978-3-030-32323-3_4

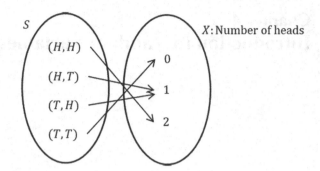

Fig. 4.1 The number of heads as a function of tossing two coins

Random variables are denoted by capital letters such as X, Y, Z. The lower-case letters represent the possible values of a random variable.

Example 1 (revisited 1) By Definition 1, X is a random variable since it is defined on the sample space S.

Definition 2 (*Discrete vs. continuous random variables*) The random variables are classified as *discrete* or *continuous* random variables. A *discrete* random variable can have a *finite* (*countable*) number of possible values. However, a *continuous* random variable can have an *infinite* number of possible values.

Example 1 (revisited 2) Since $X = \{0, 1, 2\}$, the number of the possible values of the random variable X can be counted, thus, X is a *discrete* random variable.

Example 2 If we assume that the time for waiting in a queue is a random variable that can take *any value* over the interval (0, 10) minutes, then this random variable is a *continuous* random variable, since the number of the possible values is infinite.

4.2   Basic Parameters for the Discrete and Continuous Random Variables

Every random variable is characterized by the *probability mass (density) function, cumulative distribution function, complementary cumulative distribution function (tail function), expected value, variance,* and *standard deviation.* Table 4.1 provides the formulae of the basic parameters for the discrete and continuous random variables.

Remark 1 The *probability mass function (pmf)* is defined for the *discrete* random variables, whereas the *probability density function (pdf)* is defined for the *continuous* random variables. The probability density function does *not* provide the probability of a *single value* of a continuous random variable. Rather, the area under this function for an interval of the possible values of a continuous random variable provides the probability in this interval. The probability density function can also be called as

Table 4.1 Basic parameters for the discrete and continuous random variables

Discrete random variables	Continuous random variables
1. *Probability mass function (pmf)* $P(X = i) = p(i)$ Note that $\sum_x p(x) = 1$ holds.	1. *Probability density function (pdf)* $f(x)$ $P(a \le X \le b) = \int_a^b f(x)dx$ Note that $\int_{x_{min}}^{x_{max}} f(x)dx = 1$ holds.
2. *Cumulative distribution function (cdf)* $F(i) = P(X \le i) = \sum_{x \le i} p(x)$	2. *Cumulative distribution function (cdf)* $F(i) = P(X \le i) = P(X < i) =$ $\int_{x_{min}}^{i} f(x)dx$ Note that $\frac{dF(x)}{dx} = f(x)$ holds.
3. *Complementary cumulative distribution function (tail function)* $\bar{F}(i) = 1 - F(i) = P(X > i) = \sum_{x > i} p(x)$	3. *Complementary cumulative distribution function (tail function)* $\overline{F}(i) = 1 - F(i) = P(X > i) =$ $\int_i^{x_{max}} f(x)dx$
4. *Expected value* 4.1. *Expected value of X* $E[X] = \sum_x xp(x)$ 4.2. *Expected value of g(X)* $E[g(X)] = \sum_x g(x)p(x)$ where $g(X)$ is a function of the discrete random variable X. *Special case*: $E[aX + b] = aE[X] + b$ where a, b are some constants.	4. *Expected value* 4.1. *Expected value of X* $E[X] = \int_{x_{min}}^{x_{max}} xf(x)dx$ 4.2. Expected value of $g(X)$ $E[g(X)] = \int_{x_{min}}^{x_{max}} g(x)f(x)dx$ where $g(X)$ is a function of the continuous random variable X. *Special case*: $E[aX + b] = aE[X] + b$ where a, b are some constants. *Special case*: If X is a continuous random variable that can take *only nonnegative* values, then the alternative way of calculating $E[X]$ will be $E[X] = \int_0^{x_{max}} P(X > x)dx$ $= \int_0^{x_{max}} \overline{F}(x)dx$ which is called as *integrating the tail function* method.

(continued)

Table 4.1 (continued)

Discrete random variables	Continuous random variables
5. *Variance and standard deviation*	5. *Variance and standard deviation*
5.1. *Variance of* X	5.1. *Variance of* X
$Var(X) = \sigma^2 = E(X - E[X])^2$	$Var(X) = \sigma^2 = E(X - E[X])^2$
or	or
$Var(X) = \sigma^2 = E[X^2] - (E[X])^2$	$Var(X) = \sigma^2 = E[X^2] - (E[X])^2$
Special case:	*Special case*:
$Var(aX + b) = a^2 Var(X)$	$Var(aX + b) = a^2 Var(X)$
where a is a constant.	where a is a constant.
5.2. *Standard deviation of* X	5.2. *Standard deviation of* X
$SD(X) = \sigma = \sqrt{Var(X)}$	$SD(X) = \sigma = \sqrt{Var(X)}$

density function, and the cumulative distribution function can also be called as *distribution function*. For a continuous random variable, note that $P(X \leq i) = P(X < i)$ holds since the probability in an interval is computed by integral, which means by the area under the probability density function. The *expected value* can be interpreted as the *weighted average* of the possible values of a random variable with the probabilities considered as the weights. The *variance* can be interpreted as the weighted total measure of the *deviation* of each possible value from the expected value with the probabilities considered as the weights.

Example 1 (revisited 3) The probability mass function is illustrated in Fig. 4.2.

Note that

$$\sum_x p(x) = p(0) + p(1) + p(2) = \frac{1}{4} + \frac{1}{2} + \frac{1}{4} = 1$$

The value of the *cumulative distribution function* at value 1 will be

$$F(1) = P(X \leq 1) = \sum_{x \leq 1} p(x) = p(0) + p(1) = \frac{1}{4} + \frac{1}{2} = \frac{3}{4}$$

Fig. 4.2 Probability mass function for tossing two coins experiment

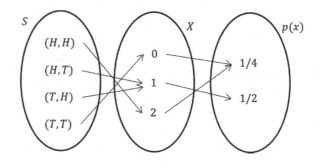

The value of the *complementary cumulative distribution function* at value 1 will be

$$\overline{F}(1) = P(X > 1) = P(X = 2) = p(2) = \frac{1}{4}$$

The *expected value* of X can be calculated as

$$E[X] = \sum_x xp(x) = (0)\left(\frac{1}{4}\right) + (1)\left(\frac{1}{2}\right) + (2)\left(\frac{1}{4}\right) = 1$$

As a special case,

$$E[2X + 5] = 2E[X] + 5 = 7$$

The *variance* of X can be calculated as

$$E[X^2] = (0)^2\left(\frac{1}{4}\right) + (1)^2\left(\frac{1}{2}\right) + (2)^2\left(\frac{1}{4}\right) = \frac{3}{2}$$

$$Var(X) = \sigma^2 = E[X^2] - (E[X])^2 = \frac{3}{2} - (1)^2 = \frac{1}{2}$$

As a special case,

$$Var(3X + 4) = (3)^2 Var(X) = 9Var(X) = \frac{9}{2}$$

The *standard deviation* of X can be calculated as

$$SD(X) = \sigma = \sqrt{Var(X)} = \sqrt{\frac{1}{2}}$$

Example 3 Consider the following *probability density function* for a continuous random variable X:

$$f(x) = \begin{cases} 2x & \text{if } 0 \leq x \leq 1 \\ 0 & \text{otherwise} \end{cases}$$

The *cumulative distribution function* of X will be

$$F(i) = P(X \leq i) = P(X < i) = \int_0^i f(x)dx = \int_0^i 2xdx = i^2$$

or

$$F(x) = x^2$$

The *complementary cumulative distribution function* of X will be

$$\overline{F}(i) = P(X > i) = 1 - F(i) = 1 - i^2$$

or

$$\overline{F}(x) = 1 - x^2$$

The *expected value* of X will be

$$E[X] = \int_{x_{min}}^{x_{max}} xf(x)dx = \int_0^1 x2xdx = \int_0^1 2x^2dx = \frac{2}{3}$$

Since X is a continuous random variable that can take *only nonnegative* values, the expected value can also be calculated by *integrating the tail function method* as follows:

$$E[X] = \int_0^1 P(X > x)dx = \int_0^1 \overline{F}(x)dx = \int_0^1 (1 - x^2)dx = \frac{2}{3}$$

The *variance* of X can be calculated as

$$E[X^2] = \int_{x_{min}}^{x_{max}} x^2 f(x)dx = \int_0^1 x^2 2xdx = \int_0^1 2x^3dx = \frac{1}{2}$$

$$Var(X) = \sigma^2 = E[X^2] - (E[X])^2 = \frac{1}{2} - \left(\frac{2}{3}\right)^2 = \frac{1}{18}$$

The *standard deviation* of X can be calculated as

$$SD(X) = \sigma = \sqrt{Var(X)} = \sqrt{\frac{1}{18}}$$

Problems

1. Please define the following random variables as *discrete* or *continuous* random variables.

    (a) Number of the typographical errors on a page of a book

   (b) Length of pencils that may vary between 18 and 22 cm
   (c) Lifetime of a machine
   (d) Number of heads if we toss a coin 10 times

Answer

(a) Since the number of the typographical errors on a page of a book is limited to the number of the letters on that page, it is *countable*, thus, this random variable is a *discrete* random variable.
(b) Since the length of pencils can take any real value between 18 and 22 cm, the number of the possible values is *not countable*, thus, this random variable is a *continuous* random variable.
(c) The number of the possible values for the lifetime of a machine is *not countable*, thus, this random variable is a *continuous* random variable.
(d) Since the number of heads can be any integer up to 10 if we toss a coin 10 times, it is *countable*, and this random variable is a *discrete* random variable.
2. Let I_A be a random variable defined for an event A such that

$$I_A = \begin{cases} 1 & \text{if } A \text{ occurs} \\ 0 & \text{otherwise} \end{cases}$$

What is the expected value and the variance of I_A?

Solution
Let
   $P(A)$: Probability that event A occurs
   Hence, the expected value of I_A will be

$$E[I_A] = (1)P(A) + (0)P(A^C) = P(A)$$

and the variance of I_A will be

$$Var(I_A) = E[I_A^2] - (E[I_A])^2$$

where

$$E[I_A^2] = (1^2)P(A) + (0^2)P(A^C) = P(A)$$

Finally,

$$Var(I_A) = P(A) - (P(A))^2 = P(A)(1 - P(A))$$

Remark 2 I_A defined for an event A is a special random variable called as *indicator variable*.

3. Two fair dice are rolled. Let X be the absolute value of "the outcome of the first die minus the outcome of the second die". Find the probability mass function of X. Calculate the probability that at least $\{X \geq 5\}$ or $\{X \leq 1\}$ will occur. Find also $E[X]$ and $Var(X)$.

Solution
The probability mass function of X will be as follows:

$$P(X = 0) = p(0) = \frac{6}{36}$$
$$P(X = 1) = p(1) = \frac{10}{36}$$
$$P(X = 2) = p(2) = \frac{8}{36}$$
$$P(X = 3) = p(3) = \frac{6}{36}$$
$$P(X = 4) = p(4) = \frac{4}{36}$$
$$P(X = 5) = p(5) = \frac{2}{36}$$

Note that

$$p(0) + p(1) + p(2) + p(3) + p(4) + p(5) = 1$$

The probability mass function of X can be plotted as in Fig. 4.3

Fig. 4.3 Probability mass function for rolling two fair dice experiment

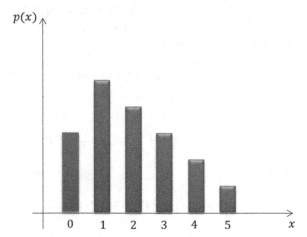

The probability that at least $\{X \geq 5\}$ or $\{X \leq 1\}$ will occur will be

$$P(\{X \geq 5\} \cup \{X \leq 1\}) = P(X \geq 5) + P(X \leq 1)$$

by Axiom 3 since $\{X \geq 5\}$ and $\{X \leq 1\}$ are *mutually exclusive* events, and

$$P(X \geq 5) + P(X \leq 1) = P(X = 5) + P(X = 0) + P(X = 1)$$
$$= \frac{2}{36} + \frac{6}{36} + \frac{10}{36} = \frac{18}{36} = \frac{1}{2}$$

The *expected value* of X will be

$$E[X] = (0)\left(\frac{6}{36}\right) + (1)\left(\frac{10}{36}\right) + (2)\left(\frac{8}{36}\right) + (3)\left(\frac{6}{36}\right) + (4)\left(\frac{4}{36}\right) + (5)\left(\frac{2}{36}\right) = \frac{70}{36} \cong 1.9444$$

and the *variance* of X will be

$$Var(X) = \sigma^2 = E[X^2] - (E[X])^2$$

where

$$E[X^2] = (0)^2\left(\frac{6}{36}\right) + (1)^2\left(\frac{10}{36}\right) + (2)^2\left(\frac{8}{36}\right) + (3)^2\left(\frac{6}{36}\right) + (4)^2\left(\frac{4}{36}\right) + (5)^2\left(\frac{2}{36}\right) = \frac{210}{36} \cong 5.8333$$

Finally,

$$Var(X) = \sigma^2 = E[X^2] - (E[X])^2 = 5.8333 - (1.9444)^2 \cong 2.0526$$

4. Let X be a continuous random variable and let

$$f(x) = \begin{cases} 2e^{-2x} & \text{if } x \geq 0 \\ 0 & \text{if } x < 0 \end{cases}$$

Verify that $f(x)$ can be a probability density function of X. Compute the expected value of X by *integrating the tail function* method, and find the variance of X. Additionally, compute $P(X > 10)$ and $P(X < 5)$.

Solution
Since

$$\int_0^\infty 2e^{-2x}dx = 1$$

$f(x)$ can be a probability density function of X. $E[X]$ can be computed by *integrating the tail function* method by using

$$E[X] = \int_0^\infty \overline{F}(x)dx$$

Since

$$F(x) = \int_0^x 2e^{-2x}dx = -e^{-2x} + 1$$

$\overline{F}(x) = 1 - F(x) = 1 - (-e^{-2x} + 1) = e^{-2x}$ holds, and

$$E[X] = \int_0^\infty e^{-2x}dx = \frac{1}{2}$$

The *variance* of X will be

$$Var(X) = \sigma^2 = E[X^2] - (E[X])^2$$

where

$$E[X^2] = \int_0^\infty x^2 2e^{-2x}dx = \frac{1}{2}$$

Finally,

$$Var(X) = \sigma^2 = E[X^2] - (E[X])^2 = \frac{1}{2} - \left(\frac{1}{2}\right)^2 = \frac{1}{4}$$

Additionally,

$$P(X > 10) = \int_{10}^\infty 2e^{-2x}dx = e^{-20} = 2.06 \times 10^{-9}$$

$$P(X < 5) = \int_0^5 2e^{-2x}dx = -e^{-10} + 1 = 0.9999$$

Warning For any continuous random variable X, note the difference between

$$\int\limits_{x_{min}}^{x_{max}} f(x)dx = 1$$

and

$$F(x) = \int\limits_{x_{min}}^{x} f(x)dx$$

5. Consider a newsboy who wants to decide the daily number of newspapers Q to purchase from a vendor. Daily demand D is a random variable, which can be considered as *discrete* or *continuous*. The unit purchasing cost of a newspaper from a vendor is c, the selling price of a newspaper is p, and the salvage value of each unsold newspaper is s. Formulate the *expected profit* for this problem.

Solution

Let

$\pi(Q)$: Profit when the newsboy purchases Q newspapers from a vendor which can be formulated as

$$\pi(Q) = p\,min(Q, D) + s(Q - D)^+ - cQ$$

where

$$(Q - D)^+ = max(Q - D, 0)$$

Hence, the expected profit will be

$$E[\pi(Q)] = pE[min(Q, D)] + sE\big[(Q - D)^+\big] - cQ$$

Remark 3 The problem of finding the optimum number of newspapers to purchase daily from a vendor with the objective of maximizing the expected daily profit is a special *single-period inventory problem* called as *newsboy (newsvendor) problem*, in which the product is a *perishable* product. Note that the perishable products are those products which have specified lifetimes after which they are considered to be inappropriate for utilization.

6. Please recall from Problem 4 in Chap. 2 that there are 5 red balls, 7 yellow balls, and 3 white balls in the urn, and 3 balls are randomly selected from this urn. Suppose that for each *red ball* withdrawn, we get -1 point; for each *yellow ball* withdrawn, we get 0 point; and for each *white ball* withdrawn, we get 1 point. Find the *expected value* of the total points.

Solution

The probability values for all possible events will be as follows:

$$P(\{R = 3\}) = \frac{\binom{5}{3}}{\binom{15}{3}} = \frac{10}{455}$$

$$P(\{Y = 3\}) = \frac{\binom{7}{3}}{\binom{15}{3}} = \frac{35}{455}$$

$$P(\{W = 3\}) = \frac{\binom{3}{3}}{\binom{15}{3}} = \frac{1}{455}$$

$$P(\{R = 1\} \cap \{Y = 1\} \cap \{W = 1\}) = \frac{\binom{5}{1}\binom{7}{1}\binom{3}{1}}{\binom{15}{3}} = \frac{105}{455}$$

$$P(\{R = 1\} \cap \{Y = 2\}) = \frac{\binom{5}{1}\binom{7}{2}}{\binom{15}{3}} = \frac{105}{455}$$

$$P(\{R = 1\} \cap \{W = 2\}) = \frac{\binom{5}{1}\binom{3}{2}}{\binom{15}{3}} = \frac{15}{455}$$

$$P(\{Y = 1\} \cap \{R = 2\}) = \frac{\binom{7}{1}\binom{5}{2}}{\binom{15}{3}} = \frac{70}{455}$$

$$P(\{Y = 1\} \cap \{W = 2\}) = \frac{\binom{7}{1}\binom{3}{2}}{\binom{15}{3}} = \frac{21}{455}$$

$$P(\{W = 1\} \cap \{R = 2\}) = \frac{\binom{3}{1}\binom{5}{2}}{\binom{15}{3}} = \frac{30}{455}$$

$$P(\{W = 1\} \cap \{Y = 2\}) = \frac{\binom{3}{1}\binom{7}{2}}{\binom{15}{3}} = \frac{63}{455}$$

Table 4.2 summarizes all possible events with their total points, and probability values.

Finally, the *expected value* of the total points will be

$$E[X] = -3\left(\frac{10}{455}\right) - 2\left(\frac{70}{455}\right) - 1\left(\frac{105}{455} + \frac{30}{455}\right) + 0\left(\frac{35}{455} + \frac{105}{455}\right)$$
$$+ 1\left(\frac{15}{455} + \frac{63}{455}\right) + 2\left(\frac{21}{455}\right) + 3\left(\frac{1}{455}\right) = -\frac{182}{455} = -0.4$$

7. $E[X] = 3$ and $Var(X) = 8$ are given for a random variable X.

(a) Find $E[(5 + X)^2]$.
(b) Find $Var(4X + 7)$.

Solution

(a)
$$E[(5 + X)^2] = E[25 + 10X + X^2]$$
$$= 25 + 10E[X] + E[X^2]$$

Table 4.2 Events, total points, and probability values for problem 6

Event	Total point	Probability
$\{R = 3\}$	-3	$\frac{10}{455}$
$\{Y = 3\}$	0	$\frac{35}{455}$
$\{W = 3\}$	3	$\frac{1}{455}$
$\{R = 1\} \cap \{Y = 1\} \cap \{W = 1\}$	0	$\frac{105}{455}$
$\{R = 1\} \cap \{Y = 2\}$	-1	$\frac{105}{455}$
$\{R = 1\} \cap \{W = 2\}$	1	$\frac{15}{455}$
$\{Y = 1\} \cap \{R = 2\}$	-2	$\frac{70}{455}$
$\{Y = 1\} \cap \{W = 2\}$	2	$\frac{21}{455}$
$\{W = 1\} \cap \{R = 2\}$	-1	$\frac{30}{455}$
$\{W = 1\} \cap \{Y = 2\}$	1	$\frac{63}{455}$

Since

$$Var(X) = \sigma^2 = E[X^2] - (E[X])^2$$

it follows that

$$E[X^2] = Var(X) + (E[X])^2$$
$$= 8 + (3)^2 = 17$$

Finally,

$$E[(5 + X)^2] = 25 + 10E[X] + E[X^2] = 25 + (10)(3) + 17 = 72$$

(b) $Var(4X + 7) = (4)^2 Var(X) = 16 Var(X) = (16)(8) = 128$

8. A shop manager wants to decide the stock quantity Q of a particular product. There is a profit of \$10 for each product sold, and there is a loss of \$2 for each product unsold. The demand for this product is a *discrete* random variable X having possible values 1,000, 750, 500 with respective probabilities $\frac{1}{6}, \frac{1}{2}, \frac{1}{3}$. Determine the *expected profit formula* as a function of Q. Compute the expected profit for $Q = 1,000$.

Solution

Let

$\pi(Q)$: Profit when stock quantity is Q

which can be formulated as

$$\pi(Q) = \begin{cases} 10X - 2(Q - X), & \text{if } X \le Q \\ 10Q, & \text{if } X > Q \end{cases}$$

Hence, the expected profit formula as a function of Q will be

$$E[\pi(Q)] = \sum_{k=0}^{Q}(10k - 2(Q - k))p(k) + \sum_{k=Q+1}^{\infty} 10Q \, p(k)$$

$$= \sum_{k=0}^{Q}(10k + 2(k - Q))p(k) + \left(\sum_{k=0}^{\infty} 10Qp(k) - \sum_{k=0}^{Q} 10Q \, p(k)\right)$$

$$= \left(\sum_{k=0}^{Q}(10k - 10Q + 2(k - Q))p(k)\right) + \left(10Q \sum_{k=0}^{\infty} p(k)\right)$$

$$= \left(\sum_{k=0}^{Q}(10(k - Q) + 2(k - Q))p(k)\right) + 10Q$$

$$= \left(\sum_{k=0}^{Q} (12(k - Q))p(k) \right) + 10Q$$

Accordingly, the expected profit for $Q = 1,000$ can be calculated as

$$E[\pi(1,000)] = 12\left[(1,000 - 1,000)\left(\frac{1}{6}\right) + (750 - 1,000)\left(\frac{1}{2}\right) + (500 - 1,000)\left(\frac{1}{3}\right) \right]$$

$$+ 10(1,000) = 12\left[-125 - \frac{500}{3} \right] + 10,000 = \$6,500$$

Chapter 5
Discrete Random Variables

Abstract This chapter introduced some special discrete random variables, i.e. binomial random variable and Bernoulli random variable, Poisson random variable, negative binomial random variable, geometric random variable, and hypergeometric random variable. The formulae for the basic parameters of these special discrete random variables have been provided. In addition to the typical examples and problems such as the number of the typographical errors on a book page, and the ball withdrawal from an urn without replacement, one special problem that is related to *acceptance sampling* in *quality control* has also been presented.

After presenting the fundamentals of the random variables in Chap. 4, this chapter will introduce special discrete random variables such as *binomial random variable* and *Poisson random variable*, which can represent several random phenomena. In this chapter, the importance of the interrelations between these special discrete random variables will be highlighted. The formulations for the basic parameters given in Chap. 4 will be still relevant, and they will be adapted for each special discrete random variable. The significance of this chapter will be clear not only in Chap. 7, which will introduce other selected topics in probability; but also in some chapters related to the stochastic processes, especially those related to *Poisson process*. Thus, several examples and problems will be provided to show how to distinguish between the discrete random variables, and how to make the calculations.

5.1 Special Discrete Random Variables

Example 1 Suppose that we toss a coin with *heads* probability $p = \frac{1}{3}$ for $n = 10$ times *independently* from each other.

Definition 1 (*Binomial random variable*) For a random experiment with *two* possible outcomes and n *independent* trials, X random variable that represents the number of the independent trials that result in *Outcome* 1 can be defined as a *binomial random variable* as illustrated in Fig. 5.1.

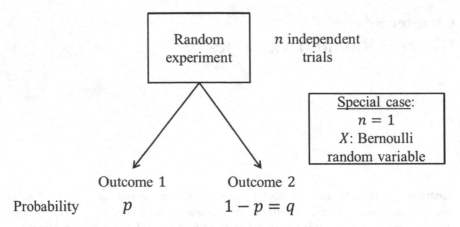

Fig. 5.1 Schematic representation of a binomial random variable

A binomial random variable X is characterized by the parameters n and p, and can be denoted by using the following notation:

$$X \sim Bin(n, p)$$

Example 1 (revisited 1) The random variable X that represents the number of the independent trials that result in *heads* can be defined as a binomial random variable, and can be denoted by

$$X \sim Bin\left(10, \frac{1}{3}\right)$$

Definition 2 (*Poisson random variable*) A Poisson random variable can be defined as an *approximation* to a *binomial random variable* $X \sim Bin(n, p)$, when n is *large* and p is *small*, as illustrated in Fig. 5.2.

A Poisson random variable X is characterized by the parameter $\lambda = np$, and can be denoted by using the following notation:

$$X \sim Poiss(\lambda)$$

Remark 1 Although the conditions $n \geq 30$, $p \leq 0.1$ have been added as a general rule for the applicability of a Poisson random variable as an *approximation* to a binomial random variable, even in cases without these conditions, the results are rather close to each other.

Example 1 (revisited 2) When we change the parameters as $n = 200$ and $p = 0.01$, $\lambda = 200(0.01) = 2$, and the random variable X that represents the number of the independent trials that result in *heads* can also be defined as a *Poisson random variable*, and can be denoted by

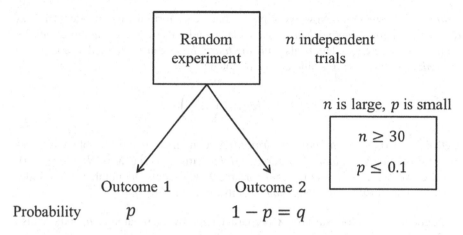

Fig. 5.2 Schematic representation of a Poisson random variable

$$X \sim Poiss(2)$$

Definition 3 (*Negative binomial random variable*) For a random experiment with *two* possible outcomes, X random variable that represents the number of the independent trials *until* the rth *Outcome* 1 can be defined as a *negative binomial random variable* as illustrated in Fig. 5.3.

A negative binomial random variable X is characterized by the parameters r and p, and can be denoted by using the following notation:

$$X \sim NegBin(r, p)$$

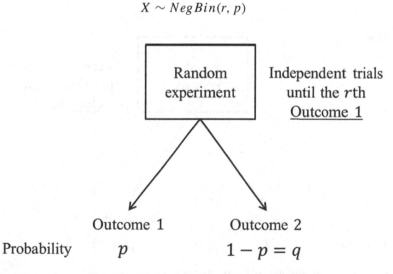

Fig. 5.3 Schematic representation of a negative binomial random variable

Example 1 (revisited 3) Now, we suppose that we perform the independent trials of tossing a coin *until* the *fourth head.* Hence, the random variable X that represents the number of the independent trials *until* the *fourth head* can be defined as a *negative binomial random variable*, and can be denoted by

$$X \sim NegBin\left(4, \frac{1}{3}\right)$$

Definition 4 (*Geometric random variable*) A geometric random variable X is a special case of a *negative binomial random variable* with $r = 1$, i.e. $X \sim NegBin(1, p)$. In other words, a geometric random variable X represents the number of the independent trials *until* the *first Outcome* 1 as illustrated in Fig. 5.4.

A geometric random variable X is characterized by the parameter p, and can also be denoted by using the following notation:

$$X \sim Geom(p)$$

Example 1 (revisited 4) Now, we suppose that we perform the independent trials of tossing a coin *until* the *first head.* Hence, the random variable X that represents the number of the independent trials *until* the *first head* can be defined as a *geometric random* variable, and can be denoted by

$$X \sim Geom\left(\frac{1}{3}\right)$$

or

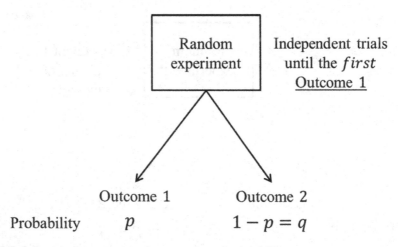

Fig. 5.4 Schematic representation of a geometric random variable

$$X \sim NegBin\left(1, \frac{1}{3}\right)$$

Definition 5 (*Hypergeometric random variable*) In a set of N elements, we suppose that m elements have a *defined property*, while the remaining $N - m$ elements do *not* have this property. If we assume that we randomly select n elements from this set *without replacement*, then the random variable X that represents the number of elements *having the defined property* can be defined as a *hypergeometric random variable*.

A hypergeometric random variable X is characterized by the parameters n, N, m, and can be denoted by using the following notation:

$$X \sim Hypergeom(n, N, m)$$

Example 2 We suppose that an urn contains 10 red balls and 5 yellow balls, and we randomly choose 4 balls from this urn *without replacement*. Accordingly, the random variable X that represents the number of the *red balls* withdrawn can be defined as a *hypergeometric random variable*, and can be denoted by

$$X \sim Hypergeom(4, 15, 10)$$

5.2 Basic Parameters for Special Discrete Random Variables

Table 5.1 provides the formulae of the basic parameters for special discrete random variables including the probability mass function, cumulative distribution function, expected value, and variance.

Example 1 (revisited 5) When we define X as a random variable that denotes the number of the independent trials that result in *heads*,

$$X \sim Bin\left(10, \frac{1}{3}\right)$$

As an example, the probability of obtaining *four heads* will be

$$P(X = 4) = p(4) = \binom{10}{4}\left(\frac{1}{3}\right)^4\left(1 - \frac{1}{3}\right)^{10-4}$$

$$= \binom{10}{4}\left(\frac{1}{3}\right)^4\left(\frac{2}{3}\right)^6 = 0.2276$$

and the expected value and the variance of X can be computed as

$$E[X] = 10\left(\frac{1}{3}\right) = \frac{10}{3} = 3.3333$$

$$Var(X) = 10\left(\frac{1}{3}\right)\left(1 - \frac{1}{3}\right) = 2.2222$$

When we change the parameters as $n = 200$ and $p = 0.01$, $\lambda = 200(0.01) = 2$, and

$$X \sim Poiss(2)$$

As an example, the probability of obtaining *four heads* will be

$$P(X = 4) = p(4) = \frac{e^{-2}(2)^4}{4!} = 0.0902$$

Note that the expected value and the variance of X will be

Table 5.1 Basic parameters for special discrete random variables

Binomial random variable	Bernoulli random variable
$X \sim Bin(n, p)$	$X \sim Bin(1, p)$
$P(X = i) = p(i) = \binom{n}{i}p^i(1-p)^{n-i}$	$P(X = i) = p(i) = p^i(1-p)^{1-i}$
$i = 0, 1, 2, \ldots, n$	$i = 0, 1$
	$F(i) = P(X \le i) = \sum_{0 \le x \le i} p^x(1-p)^{1-x}$
$F(i) = P(X \le i) = \sum_{0 \le x \le i}\binom{n}{x}p^x(1-p)^{n-x}$	$i = 0, 1$
	$E[X] = p$
$i = 0, 1, 2, \ldots, n$	$Var(X) = p(1-p)$
$E[X] = np$	
$Var(X) = np(1-p)$	
Negative binomial random variable	Geometric random variable
$X \sim NegBin(r, p)$	$X \sim Geom(p)$ or $X \sim NegBin(1, p)$
$P(X = i) = p(i) = \binom{i-1}{r-1}p^r(1-p)^{i-r}$	$P(X = i) = p(i) = p(1-p)^{i-1}$
$i = r, r+1, r+2, \ldots$	$i = 1, 2, 3, \ldots$
$F(i) = P(X \le i)$	$F(i) = P(X \le i) = \sum_{1 \le x \le i} p(1-p)^{x-1}$
$= \sum_{r \le x \le i}\binom{x-1}{r-1}p^r(1-p)^{x-r}$	$i = 1, 2, 3, \ldots$
	$E[X] = \frac{1}{p}$
$i = r, r+1, r+2, \ldots$	$Var(X) = \frac{(1-p)}{p^2}$
$E[X] = \frac{r}{p}$	
$Var(X) = \frac{r(1-p)}{p^2}$	

(continued)

Table 5.1 (continued)

Poisson random variable	Hypergeometric random variable
$X \sim Poiss(\lambda)$	$X \sim Hypergeom(n, N, m)$
$P(X = i) = p(i) = \frac{e^{-\lambda}\lambda^i}{i!}$	
$i = 0, 1, 2, \ldots$	$P(X = i) = p(i) = \dfrac{\dbinom{m}{i}\dbinom{N-m}{n-i}}{\dbinom{N}{n}}$
$F(i) = P(X \le i) = \sum\limits_{0 \le x \le i} \frac{e^{-\lambda}\lambda^x}{x!}$	
$i = 0, 1, 2, \ldots$	$n - (N - m) \le i \le min(n, m)$
$E[X] = \lambda$	
$Var(X) = \lambda$	$F(i) = P(X \le i) = \sum\limits_{0 \le x \le i} \dfrac{\dbinom{m}{x}\dbinom{N-m}{n-x}}{\dbinom{N}{n}}$
	$n - (N - m) \le i \le min(n, m)$
	$E[X] = \frac{nm}{N}$
	$Var(X) = n\left(\frac{m}{N}\right)\left(1 - \frac{m}{N}\right)\left(1 - \frac{n-1}{N-1}\right)$

$$E[X] = Var(X) = \lambda = 2$$

When we define X as a random variable that denotes the number of the independent trials *until* the *fourth head*,

$$X \sim NegBin\left(4, \frac{1}{3}\right)$$

As an example, the probability that the *fourth* head appears on the *twelfth trial* will be

$$P(X = 12) = p(12) = \binom{12 - 1}{4 - 1}\left(\frac{1}{3}\right)^4\left(1 - \frac{1}{3}\right)^{12-4}$$

$$= \binom{11}{3}\left(\frac{1}{3}\right)^4\left(\frac{2}{3}\right)^8 = 0.0795$$

and the expected value and the variance of X can be computed as

$$E[X] = \frac{4}{\left(\frac{1}{3}\right)} = 12$$

$$Var(X) = \frac{4\left(1 - \frac{1}{3}\right)}{\left(\frac{1}{3}\right)^2} = 24$$

When we define X as a random variable that denotes the number of the independent trials *until* the *first head*,

$$X \sim Geom\left(\frac{1}{3}\right)$$

As an example, the probability that the *first* head appears on the *sixth* trial will be

$$P(X = 6) = p(6) = \frac{1}{3}\left(1 - \frac{1}{3}\right)^{6-1} = \frac{1}{3}\left(\frac{2}{3}\right)^{5} = 0.0439$$

and the expected value and the variance of X can be computed as

$$E[X] = \frac{1}{\left(\frac{1}{3}\right)} = 3$$

$$Var(X) = \frac{\left(1 - \frac{1}{3}\right)}{\left(\frac{1}{3}\right)^{2}} = 6$$

Example 2 (revisited 1) Recall that the urn contains 10 red balls and 5 yellow balls, we randomly choose 4 balls from this urn *without replacement*, and the random variable X represents the number of the *red balls* withdrawn. Hence,

$$X \sim Hypergeom(4, 15, 10)$$

As an example, the probability that 2 *red balls* will be withdrawn will be

$$P(X = 2) = p(2) = \frac{\binom{10}{2}\binom{15-10}{4-2}}{\binom{15}{4}} = \frac{\binom{10}{2}\binom{5}{2}}{\binom{15}{4}} = 0.3297$$

and the expected value and the variance of X will be

$$E[X] = \frac{(4)(10)}{15} = \frac{40}{15} = 2.6667$$

$$Var(X) = 4\left(\frac{10}{15}\right)\left(1 - \frac{10}{15}\right)\left(1 - \frac{(4-1)}{(15-1)}\right) = \frac{44}{63} = 0.6984$$

Problems

1. A system with *seven components* functions if *at least* its *three components* function. Each component functions with probability $\frac{1}{3}$ *independently* from each

other. What is the probability that the system functions? What is the expected value and the variance of the number of components that function? Solve the problem by assuming the random variable as a binomial random variable and as a Poisson random variable.

Solution

Let

X: Number of components that function

When we assume

$$X \sim Bin\left(7, \frac{1}{3}\right)$$

it follows that

$$P(\text{system functions}) = P(X \geq 3) = \sum_{i=3}^{7} \binom{7}{i}\left(\frac{1}{3}\right)^i \left(1 - \frac{1}{3}\right)^{7-i} = 0.4294$$

$$E[X] = 7\left(\frac{1}{3}\right) \cong 2.3333$$

$$Var(X) = 7\left(\frac{1}{3}\right)\left(1 - \frac{1}{3}\right) = \frac{14}{9} \cong 1.5556$$

When we assume

$$X \sim Poiss\left(\frac{7}{3}\right)$$

it follows that

$$P(\text{system functions}) = P(X \geq 3) = \sum_{i=3}^{7} \frac{e^{-\frac{7}{3}}\left(\frac{7}{3}\right)^i}{i!} = 0.4099$$

$$E[X] = Var(X) = \lambda = \frac{7}{3} \cong 2.3333$$

2. The number of the typographical errors on a book page is assumed to be a *Poisson random variable* with expected value $\frac{1}{5}$.

(a) What is the probability that there is *at most* 1 error on a randomly selected page?
(b) What is the probability that there are *at least* 3 errors on a randomly selected page?
(c) Given that there is *at most* 1 error on a randomly selected page, what is the probability that there is *no* error?

Solution

Let

X: Number of the typographical errors on a randomly selected page
where $E[X] = \lambda = \frac{1}{5}$.

(a) The probability that there is *at most* 1 error on a randomly selected page will be

$$P(X \leq 1) = P(X = 0) + P(X = 1) = \frac{e^{-\frac{1}{5}}\left(\frac{1}{5}\right)^0}{0!} + \frac{e^{-\frac{1}{5}}\left(\frac{1}{5}\right)^1}{1!}$$

$$= \left(\frac{6}{5}\right)e^{-\frac{1}{5}} = 0.9825$$

(b) The probability that there are *at least* 3 errors on a randomly selected page will be

$$P(X \geq 3) = 1 - [P(X = 0) + P(X = 1) + P(X = 2)]$$

$$= 1 - \left[\frac{e^{-\frac{1}{5}}\left(\frac{1}{5}\right)^0}{0!} + \frac{e^{-\frac{1}{5}}\left(\frac{1}{5}\right)^1}{1!} + \frac{e^{-\frac{1}{5}}\left(\frac{1}{5}\right)^2}{2!} \right]$$

$$= 1 - \left[\left(\frac{61}{50}\right)e^{-\frac{1}{5}} \right] = 0.0011$$

(c) Given that there is *at most* 1 error on a randomly selected page, the probability that there is *no* error will be

$$P(\{X = 0\}|\{X \leq 1\}) = \frac{P(\{X = 0\} \cap \{X \leq 1\})}{P(X \leq 1)}$$

$$= \frac{P(X = 0)}{P(X \leq 1)} = \frac{\left(\frac{e^{-\frac{1}{5}}\left(\frac{1}{5}\right)^0}{0!}\right)}{0.9825} = 0.8333$$

where $P(X \leq 1) = 0.9825$ follows from (a).

3. In an urn, there are 4 white, 7 red, and 3 black balls. Each time a ball is withdrawn, its color is noted, and put into the urn again. What is the probability that the number of the balls withdrawn *until* the *first black* ball is greater than or equal to 8? What is the expected value and the variance of the number of balls withdrawn *until* the *first black* ball?

Solution

Let

X: Number of balls withdrawn *until* the *first black* ball

By noting that the probability of obtaining a black ball is $\frac{3}{4+7+3} = \frac{3}{14}$ for each withdrawal,

$$X \sim Geom\left(\frac{3}{14}\right)$$

$$P(X \geq 8) = \sum_{i=7}^{\infty}\left(\frac{11}{14}\right)^i\left(\frac{3}{14}\right) = \left(\frac{3}{14}\right)\sum_{i=7}^{\infty}\left(\frac{11}{14}\right)^i$$

$$= \left(\frac{3}{14}\right)\left[\sum_{i=0}^{\infty}\left(\frac{11}{14}\right)^i - \sum_{i=0}^{6}\left(\frac{11}{14}\right)^i\right]$$

$$= \left(\frac{3}{14}\right)\left[\frac{1}{\left(1-\frac{11}{14}\right)} - \frac{\left(1-\left(\frac{11}{14}\right)^7\right)}{\left(1-\frac{11}{14}\right)}\right]$$

$$= \left(\frac{11}{14}\right)^7 = 0.1849$$

$$E[X] = \frac{1}{\left(\frac{3}{14}\right)} = \frac{14}{3} \cong 4.6667$$

$$Var(X) = \frac{\left(1-\frac{3}{14}\right)}{\left(\frac{3}{14}\right)^2} = \frac{154}{9} \cong 17.1111$$

4. In a batch of 200 pencils, the probability of a pencil to be defective is 0.02. You randomly sample 10 pencils from this batch for quality control, and accept the batch if the number of the defective pencils is no more than 1. What is the probability that this batch is accepted?

Solution

Let

X: Number of the defective pencils in the sample

By assuming X as a *hypergeometric* random variable such that

$$X \sim Hypergeom(10, 200, 4)$$

where the third parameter is computed as $(200)(0.02) = 4$,

$$P(\text{acceptance}) = P(X \leq 1) = \sum_{x=0}^{1} \frac{\binom{4}{x}\binom{200-4}{10-x}}{\binom{200}{10}} = 0.9872$$

By assuming X as a *binomial* random variable such that

$$X \sim Bin(10, 0.02)$$

$$P(\text{acceptance}) = P(X \leq 1) = \sum_{x=0}^{1} \binom{10}{x}(0.02)^x (1-0.02)^{10-x} = 0.9838$$

By assuming X as a *Poisson* random variable such that

$$X \sim Poiss(0.2)$$

where $\lambda = 10(0.02) = 0.2$,

$$P(\text{acceptance}) = P(X \leq 1) = \sum_{x=0}^{1} \frac{e^{-0.2}(0.2)^x}{x!} = 0.9825$$

Remark 2 This problem is a kind of *acceptance sampling* problem from *quality control* to decide the acceptance of a batch of items. *Hypergeometric* random variable is used for the assumption of *sampling without replacement*, while *binomial* and *Poisson* random variables can be used for the assumption of *sampling with replacement*. When the batch size is relatively small, hypergeometric random variable can be used; when the batch size is large, binomial and Poisson random variables can be preferred.

5. In a roulette wheel game, a person assumes that he wins when the outcome is 00 or a number between 0 and 5, including 0 and 5. If the sample space is defined as $S = \{00, 0, 1, 2, 3, \ldots, 36\}$, what is the probability that

(a) this person wins 8 times out of total 30 spins?
(b) the first win will be after the 15th spin?
(c) the third win will be at the 20th spin?

Solution

(a) Let
    X: Number of wins

Accordingly,

$$X \sim Bin\left(30, \frac{7}{38}\right)$$

and the probability that this person wins 8 times out of total 30 spins will be

$$P(X = 8) = \binom{30}{8}\left(\frac{7}{38}\right)^8\left(1 - \frac{7}{38}\right)^{30-8} = \binom{30}{8}\left(\frac{7}{38}\right)^8\left(\frac{31}{38}\right)^{22} = 0.088$$

(b) Let

X: Number of spins *until* the *first* win
Hence,

$$X \sim Geom\left(\frac{7}{38}\right)$$

and the probability that the first win will be after the 15th spin will be

$$P(X \geq 16) = \sum_{i=15}^{\infty}\left(\frac{31}{38}\right)^i\left(\frac{7}{38}\right) = \left(\frac{7}{38}\right)\sum_{i=15}^{\infty}\left(\frac{31}{38}\right)^i$$

$$= \left(\frac{7}{38}\right)\left[\sum_{i=0}^{\infty}\left(\frac{31}{38}\right)^i - \sum_{i=0}^{14}\left(\frac{31}{38}\right)^i\right]$$

$$= \left(\frac{7}{38}\right)\left[\frac{1}{\left(1 - \frac{31}{38}\right)} - \frac{\left(1 - \left(\frac{31}{38}\right)^{15}\right)}{\left(1 - \frac{31}{38}\right)}\right]$$

$$= \left(\frac{31}{38}\right)^{15} = 0.0472$$

(c) Let

X: Number of spins *until* the *third* win
Thus,

$$X \sim NegBin\left(3, \frac{7}{38}\right)$$

and the probability that the *third* win will be at the 20th spin will be

$$P(X = 20) = \binom{20-1}{3-1}\left(\frac{7}{38}\right)^3\left(1 - \frac{7}{38}\right)^{20-3}$$

$$= \binom{19}{2}\left(\frac{7}{38}\right)^{3}\left(\frac{31}{38}\right)^{17} = 0.0336$$

6. In a zoo with 250 animals, the probability of an animal to be sick on a given day is 0.05. What is the probability that the number of sick animals on a given day is at least 20 given that at least 5 animals are sick?

Solution Let
    X: Number of sick animals on a given day
    Hence,

$$X \sim Poiss(12.5)$$

where

$$\lambda = 250(0.05) = 12.5$$

The probability that the number of sick animals on a given day is at least 20 given that at least 5 animals are sick will be

$$
\begin{aligned}
P(\{X \geq 20\}|\{X \geq 5\}) &= \frac{P(\{X \geq 20\} \cap \{X \geq 5\})}{P(X \geq 5)} \\
&= \frac{P(X \geq 20)}{P(X \geq 5)} = \frac{1 - P(X \leq 19)}{1 - P(X \leq 4)} \\
&= \frac{1 - \sum_{x=0}^{19} \frac{e^{-12.5}(12.5)^{x}}{x!}}{1 - \sum_{x=0}^{4} \frac{e^{-12.5}(12.5)^{x}}{x!}} \\
&= \frac{1 - 0.9694}{1 - 0.0053} = 0.0308
\end{aligned}
$$

7. Let X be a binomial random variable with expected value 10 and variance 6. Find $P(X = 10)$.

Solution
Let

$$X \sim Bin(n, p)$$

where

$$E[X] = np = 10$$

$$Var(X) = np(1 - p) = 6$$

Accordingly,

$$\frac{Var(X)}{E[X]} = \frac{np(1-p)}{np} = 1 - p = \frac{6}{10}$$

and

$$p = 1 - \frac{6}{10} = \frac{4}{10} = 0.4$$

$$n = \frac{10}{p} = \frac{10}{0.4} = 25$$

As a result,

$$X \sim Bin(25, 0.4)$$

and

$$P(X = 10) = \binom{25}{10}(0.4)^{10}(1 - 0.4)^{25-10} = \binom{25}{10}(0.4)^{10}(0.6)^{15} = 0.1612$$

Chapter 6
Continuous Random Variables

Abstract This chapter introduced special continuous random variables including uniform random variable, exponential random variable, normal random variable, standard normal random variable, and other continuous random variables including gamma random variable, lognormal random variable, and Weibull random variable. The standard normal random variable approximation to a binomial random variable has also been provided. The formulae of the basic parameters for each continuous random variable have been given, and several typical examples and problems have been presented for each continuous random variable.

This chapter will introduce special continuous random variables. The learnings from Chap. 4 will be still relevant, and the formulations for the basic parameters from Chap. 4 will be adapted for each continuous random variable. This chapter is very important in its own sense, but the continuous random variables are also very crucial to understand the other selected topics in basic probability in Chap. 7. Two continuous random variables, i.e. *exponential random variable* and *gamma random variable* are also extremely important in some chapters related to the stochastic processes, especially those related to *Poisson process*. Since normal random variable *approximation* to a binomial random variable is possible, there is also a basic interrelation between Chaps. 5 and 6, which is illustrated with some examples and problems in this chapter.

6.1 Special Continuous Random Variables

Example 1 Suppose that the time spent for waiting in a bank queue is assumed to be a continuous random variable X.

Definition 1 (*Uniform random variable*) A continuous random variable X defined over the interval (α, β) is said to be a *uniform* random variable, if its probability density function is defined as

© Springer Nature Switzerland AG 2019

E. Bas, *Basics of Probability and Stochastic Processes*,

https://doi.org/10.1007/978-3-030-32323-3_6

Fig. 6.1 Probability density function of a uniform random variable

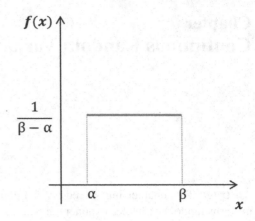

$$f(x) = \frac{1}{\beta - \alpha} \quad \alpha < x < \beta$$

which is illustrated in Fig. 6.1.

A uniform random variable X defined over the interval (α, β) is characterized by the parameters α and β, and can be denoted by using the following notation:

$$X \sim Uniform(\alpha, \beta)$$

Example 1 (revisited 1) If we assume the waiting time as a uniform random variable defined over the interval (0, 10) minutes, then

$$X \sim Uniform(0, 10)$$

and

$$f(x) = \frac{1}{10} \quad 0 < x < 10$$

which is illustrated in Fig. 6.2.

As an example, the probability that the waiting time is less than 3 min will be

$$P(X < 3) = F(3) = \int_0^3 \frac{1}{10}dx = \frac{3}{10}$$

and the probability that the waiting time is between 5 and 7 min will be

$$P(5 < X < 7) = F(7) - F(5) = \int_5^7 \frac{1}{10}dx = \frac{7}{10} - \frac{5}{10} = \frac{2}{10} = \frac{1}{5}$$

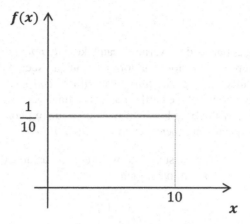

Fig. 6.2 Probability density function of uniform random variable for example 1 (revisited 1)

Definition 2 (*Exponential random variable*) A continuous random variable X is said to be an *exponential* random variable, if its probability density function is defined as

$$f(x) = \lambda e^{-\lambda x} \quad x \geq 0, \quad \lambda > 0$$

which is illustrated in Fig. 6.3.

An exponential random variable X is characterized by the parameter λ, and can be denoted by using the following notation:

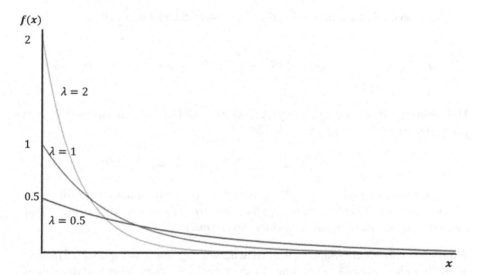

Fig. 6.3 Probability density function of an exponential random variable

$$X \sim \exp(\lambda)$$

Remark 1 An exponential random variable can take *only nonnegative* values. They are appropriate to represent various random phenomena such as the *lifetimes* and the *repair times* of machines, or the *time between two consecutive events* such as the time between two consecutive earthquakes, the time between two consecutive arrivals of customers, or the time between two consecutive phone calls. λ is a *rate* that represents the expected number of events *per unit of time*.

Example 1 (revisited 2) If we assume the waiting time as an exponential random variable X with rate $\lambda = \frac{1}{5}$ per minute, then

$$X \sim \exp\left(\frac{1}{5}\right)$$

and

$$f(x) = \frac{1}{5}e^{-\frac{1}{5}x} \quad x \geq 0$$

As an example, the probability that the waiting time is less than 3 min will be

$$P(X < 3) = F(3) = \int_{0}^{3} \frac{1}{5}e^{-\frac{1}{5}x}dx = 1 - e^{-\frac{3}{5}} = 0.4512$$

and the probability that the waiting time is between 5 and 7 min will be

$$P(5 < X < 7) = F(7) - F(5) = \int_{5}^{7} \frac{1}{5}e^{-\frac{1}{5}x}dx = e^{-1} - e^{-\frac{7}{5}} = 0.1213$$

Definition 3 (*Memoryless property*) A random variable X is said to have *memoryless property* if the following equation holds:

$$P(X > i + j | X > j) = P(X > i) \quad \text{for all } i, j \geq 0$$

Remark 2 It can be mathematically proved that *exponential* random variable is the *unique continuous* random variable, and *geometric* random variable is the *unique discrete* random variable having *memoryless property*.

Example 1 (revisited 3) Suppose again that the waiting time is an exponential random variable with rate $\lambda = \frac{1}{5}$ per minute. As an example, the probability that the waiting time will exceed 5 min given that it exceeds 2 min will be

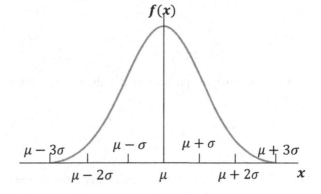

Fig. 6.4 Probability density function of a normal random variable

$$P(X > 5 | X > 2) = P(X > 3) = \overline{F}(3) = \int_{3}^{\infty} \frac{1}{5} e^{-\frac{1}{5}x} dx = e^{-\frac{3}{5}} = 0.5488$$

by *memoryless property* of exponential random variable.

Definition 4 (*Normal random variable*) A continuous random variable X is said to be a *normal* random variable, if its probability density function is defined as

$$f(x) = \frac{1}{\sigma\sqrt{2\pi}} e^{-(x-\mu)^2/2\sigma^2} \quad -\infty < x < \infty$$

which is illustrated in Fig. 6.4.

A normal random variable X is characterized by the parameters μ and σ^2, where $\mu = E[X]$ and $\sigma^2 = Var(X)$, and can be denoted by using the following notation:

$$X \sim N(\mu, \sigma^2)$$

Remark 3 Normal random variable is a very important continuous random variable that represents many random phenomena such as the heights, weights, *IQ*-levels of the randomly selected individuals in a community, and the exam grades of the students in a classroom. The probability density function of a normal random variable is always *symmetric* around μ.

Example 1 (revisited 4) If we assume the waiting time as a normal random variable with expected value $\mu = 5$ and variance $\sigma^2 = 25$, then

$$X \sim N(5, 25)$$

and

$$f(x) = \frac{1}{\sqrt{25}\sqrt{2\pi}}e^{-(x-5)^2/((2)(25))} = \frac{1}{5\sqrt{2\pi}}e^{-(x-5)^2/50} \quad \text{for } 0 < x < \infty$$

As an example, the probability that the waiting time is less than 3 min will be

$$P(X < 3) = F(3) = \int_0^3 \frac{1}{5\sqrt{2\pi}}e^{-(x-5)^2/50}dx = 0.3446$$

Remark 4 Note that it is computationally burdensome to find the probabilities by using the probability density function of a normal random variable.

Definition 5 (*68%, 95%, 99.7% rule*) If

$$X \sim N(\mu, \sigma^2)$$

then it can be proved that

$$P(\mu - \sigma \le X \le \mu + \sigma) \cong 68\%$$

$$P(\mu - 2\sigma \le X \le \mu + 2\sigma) \cong 95\%$$

$$P(\mu - 3\sigma \le X \le \mu + 3\sigma) \cong 99.7\%$$

and this is called as 68%, 95%, 99.7% rule, which is illustrated in Fig. 6.5.

Definition 6 (*Standard normal random variable*) If

$$X \sim N(\mu, \sigma^2)$$

Fig. 6.5 68%, 95%, 99.7% rule for a normal random variable

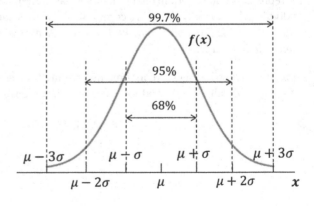

Fig. 6.6 Probability density function of a standard normal random variable

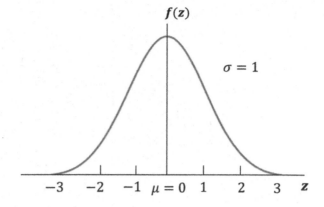

then

$$Z = \frac{X - \mu}{\sigma}$$

is said to be a *standard (unit) normal* random variable, where

$$Z \sim N(0, 1)$$

Note that a *standard (unit)* normal random variable is a *special normal* random variable with parameters $\mu - 0$ and $\sigma^2 = 1$.

The probability density function of a *standard (unit)* normal random variable Z will be

$$f(z) = \frac{1}{\sqrt{2\pi}} e^{-z^2/2} \quad -\infty < z < \infty$$

which is illustrated in Fig. 6.6.

Remark 5 The probability density function of a standard normal random variable is also *symmetric* around the parameter $\mu = 0$. Because of the symmetricity of the probability density function of a standard normal random variable Z, the following equations hold:

$$P(Z \leq -z) = P(Z \geq z)$$
$$P(Z \leq z) = P(Z \geq -z)$$

It is customary to write

$$\Phi(z) = P(Z \leq z)$$

as the cumulative distribution function of a standard normal random variable Z, and

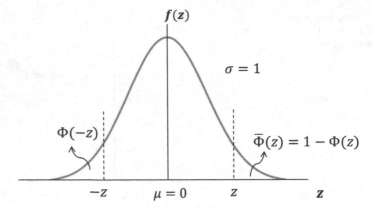

Fig. 6.7 Representation of the equation $\Phi(-z) = \overline{\Phi}(z) = 1 - \Phi(z)$

$$\overline{\Phi}(z) = P(Z > z)$$

as the complementary cumulative distribution function (tail function) of a standard normal random variable Z. Accordingly, by considering the abovementioned equations,

$$\Phi(-z) = \overline{\Phi}(z) = 1 - \Phi(z)$$
$$\Phi(z) = \overline{\Phi}(-z) = 1 - \Phi(-z)$$

will hold. Figure 6.7 shows the representation of the equation $\Phi(-z) = \overline{\Phi}(z) = 1 - \Phi(z)$.

Example 1 (revisited 5) If we assume again the waiting time as a normal random variable with expected value $\mu = 5$ and variance $\sigma^2 = 25$, then

$$X \sim N(5, 25)$$

As an example, the probability that the waiting time is less than 3 min will be

$$P(X < 3) = F(3) = P\left(\frac{X - \mu}{\sigma} < \frac{3 - \mu}{\sigma}\right) = P\left(\frac{X - 5}{\sqrt{25}} < \frac{3 - 5}{\sqrt{25}}\right)$$
$$= P(Z < -0.4) = \Phi(-0.4) = 1 - \Phi(0.4) = 1 - 0.6554 = 0.3446$$

and the probability that the waiting time is between 5 and 7 min will be

$$P(5 < X < 7) = F(7) - F(5) = P\left(\frac{5 - 5}{\sqrt{25}} < \frac{X - 5}{\sqrt{25}} < \frac{7 - 5}{\sqrt{25}}\right)$$
$$= P(0 < Z < 0.4) = \Phi(0.4) - \Phi(0) = 0.6554 - 0.5 = 0.1554$$

Example 2 The proof of $P(\mu - \sigma \leq X \leq \mu + \sigma) \cong 68\%$ from $68\%, 95\%, 99.7\%$ rule will be as follows:

$$P(\mu - \sigma \leq X \leq \mu + \sigma) = P\left(\frac{\mu - \sigma - \mu}{\sigma} \leq \frac{X - \mu}{\sigma} \leq \frac{\mu + \sigma - \mu}{\sigma}\right)$$
$$= P(-1 \leq Z \leq 1) = \Phi(1) - \Phi(-1)$$
$$= \Phi(1) - (1 - \Phi(1)) = 2\Phi(1) - 1$$
$$= 2(0.8413) - 1 = 0.6826 \cong 68\%$$

Definition 7 (*Standard normal random variable approximation to a binomial random variable*) A *binomial* random variable $X \sim Bin(n, p)$ can be *approximated* to a *standard normal* random variable by

$$Z = \frac{X - \mu}{\sigma} = \frac{X - np}{\sqrt{np(1 - p)}}$$

where $Z \sim N(0, 1)$.

By considering so-called *continuity correction* for the approximation,

$$P(a \leq X \leq b) \cong P(a - 0.5 \leq X \leq b + 0.5) = P\left(\frac{a - 0.5 - np}{\sqrt{np(1 - p)}} \leq Z \leq \frac{b + 0.5 - np}{\sqrt{np(1 - p)}}\right)$$
$$= \Phi\left(\frac{b + 0.5 - np}{\sqrt{np(1 - p)}}\right) - \Phi\left(\frac{a - 0.5 - np}{\sqrt{np(1 - p)}}\right)$$

and

$$P(X = a) \cong P(a - 0.5 \leq X \leq a + 0.5) = P\left(\frac{a - 0.5 - np}{\sqrt{np(1 - p)}} \leq Z \leq \frac{a + 0.5 - np}{\sqrt{np(1 - p)}}\right)$$
$$= \Phi\left(\frac{a + 0.5 - np}{\sqrt{np(1 - p)}}\right) - \Phi\left(\frac{a - 0.5 - np}{\sqrt{np(1 - p)}}\right)$$

Figure 6.8 denotes the representation of a standard normal random variable

Fig. 6.8 Standard normal random variable approximation to a binomial random variable

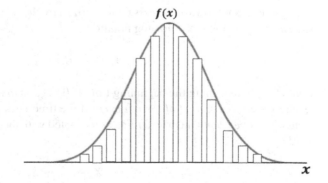

approximation to a binomial random variable.

Remark 6 As n gets larger, the probability mass function of a *binomial* random variable *approximates* to the probability density function of a *normal* random variable. It should be noted that we make the *continuity correction* by using $+0.5$ and/or -0.5.

Example 3 A coin with heads probability $p = \frac{1}{4}$ is tossed for 50 times *independently* from each other. The random variable X is defined as the number of the independent trials that result in *heads*.

(a) The probability of $X = 18$ can be computed by assuming X as a *binomial* random variable such that

$$P(X = 18) = p(18) = \binom{50}{18}\left(\frac{1}{4}\right)^{18}\left(1 - \frac{1}{4}\right)^{50-18} = \binom{50}{18}\left(\frac{1}{4}\right)^{18}\left(\frac{3}{4}\right)^{32} = 0.0264$$

(b) The probability of $X = 18$ can also be computed by assuming a standard normal random variable *approximation* to a binomial random variable X such that

$$P(X = 18) \cong P(17.5 \le X \le 18.5) = P\left(\frac{17.5 - (50)\left(\frac{1}{4}\right)}{\sqrt{50\left(\frac{1}{4}\right)\left(1 - \frac{1}{4}\right)}} \le Z \le \frac{18.5 - (50)\left(\frac{1}{4}\right)}{\sqrt{50\left(\frac{1}{4}\right)\left(1 - \frac{1}{4}\right)}}\right)$$

$$= P\left(\frac{17.5 - 12.5}{\sqrt{9.375}} \le Z \le \frac{18.5 - 12.5}{\sqrt{9.375}}\right)$$

$$= P(1.6330 \le Z \le 1.9596)$$

$$= \Phi(1.9596) - \Phi(1.6330)$$

$$\cong 0.975 - 0.9484 = 0.0266$$

Definition 8 (*Gamma random variable*) A continuous random variable t_n is said to be a *gamma* random variable, if its probability density function is defined as

$$f_{t_n}(t) = \frac{\lambda e^{-\lambda t}(\lambda t)^{n-1}}{(n - 1)!}, t \ge 0$$

A gamma random variable t_n is characterized by the parameters n and λ, and can be denoted by using the following notation:

$$t_n \sim Gamma(n, \lambda)$$

Remark 7 By assuming the beginning time as 0, a *gamma* random variable can also be interpreted as the *time of the nth event* if the time between any two consecutive events is *independent* and *exponentially* distributed with the same parameter. In other words,

$$t_n = X_1 + X_2 + \cdots + X_n$$

is a *gamma* random variable, where

$$X_i \sim \exp(\lambda) \quad i = 1, 2, \ldots, n$$

are *independent*, and X_i represents the time between the $(i-1)$st and the ith event.

Example 4 The time between the arrivals of *any* two consecutive customers to a post office is assumed to be an *independent exponential* random variable with parameter $\lambda = \frac{1}{6}$ per minute. Hence, the arrival time of the nth customer t_n will be a *gamma* random variable.

As an example, the arrival time of the 10th customer can be denoted as

$$t_{10} \sim Gamma\left(10, \frac{1}{6}\right)$$

and the probability that the arrival time of the 10th customer exceeds 45 min can be computed as

$$P(t_{10} > 45) = \int_{45}^{\infty} \frac{\left(\frac{1}{6}\right)e^{-\frac{1}{6}t}\left(\frac{1}{6}t\right)^{10-1}}{(10-1)!} dt$$

$$= \int_{45}^{\infty} \frac{\left(\frac{1}{6}\right)e^{-\frac{1}{6}t}\left(\frac{1}{6}t\right)^{9}}{9!} dt = 0.7762$$

Definition 9 (*Lognormal random variable*) A continuous *nonnegative* random variable Y is said to be a *lognormal* random variable if

$$Y = e^X$$

where

$$X \sim N(\mu, \sigma^2)$$

and if its probability density function is defined as

$$f(y) = \frac{1}{y\sigma\sqrt{2\pi}} e^{-(\ln y - \mu)^2/2\sigma^2}, \ y \geq 0$$

A lognormal random variable Y is characterized by the parameters μ and σ^2, and can be denoted by using the following notation:

$$Y \sim lognorm(\mu, \sigma^2)$$

Example 5 A lognormal random variable Y is assumed to model the stock prices of a company such that

$$Y = e^X$$

where

$$X \sim N(3, 5)$$

As an example, the probability that the stock price of the company exceeds \$2 will be

$$P(Y > 2) = \int_{2}^{\infty} \frac{1}{y\sqrt{5}\sqrt{2\pi}} e^{-(lny-3)^2/10} dy = 0.8489$$

As an alternative solution,

$$P(Y > 2) = P\left(e^X > 2\right) = P(X > \ln(2))$$
$$= P\left(\frac{X-3}{\sqrt{5}} > \frac{\ln(2)-3}{\sqrt{5}}\right) = P(Z > -1.0317)$$
$$= P(Z \leq 1.0317) = \Phi(1.0317) \cong 0.8485$$

Remark 8 Lognormal random variable is commonly used to model the *stock prices* of companies. This random variable is also very important in *Reliability Engineering* for the analysis of the *fatigue failures* and the *down times* of the technical systems.

Definition 10 (*Weibull random variable*) A continuous random variable X is said to be a *Weibull* random variable if its probability density function is defined as

$$f(x) = \alpha \lambda^\alpha x^{\alpha-1} e^{-(\lambda x)^\alpha} x \geq 0$$

where

$$\alpha : \text{Shape parameter}$$

$$\lambda : \text{Scale parameter}$$

A Weibull random variable X is characterized by the parameters α, λ, and can be denoted by using the following notation:

$$X \sim Weib(\alpha, \lambda)$$

Remark 9 Weibull random variable is widely used in different engineering areas. It is especially proper for modelling the *lifetime of the weakest link* in the technical systems. Note that for $\alpha = 1$, a *Weibull* random variable is equal to an *exponential* random variable.

Example 6 Let X be a Weibull random variable that denotes the amount of emissions from a randomly selected engine. Assume that $\alpha = 2$ and $\lambda = \frac{1}{10}$. The probability that X is greater than 15 will be

$$P(X > 15) = \int_{15}^{\infty} 2\left(\frac{1}{10}\right)^2 x^{2-1} e^{-\left(\frac{1}{10}x\right)^2} dx = \int_{15}^{\infty} 0.02\, x\, e^{-\left(\frac{1}{100}x^2\right)} dx = 0.1054$$

6.2 Basic Parameters for Special Continuous Random Variables

Table 6.1 provides the formulae of the basic parameters for special continuous random variables.

Problems

1. The position X of a particle is *uniformly* distributed over the interval $(-3, 4)$. What is the probability that the position of the particle will be nonnegative? What is the expected value and the variance of the position of the particle? Determine the probability density function of $|X|$.

Solution
Let

$$X : \text{Position of the particle}$$

Since

$$X \sim Uniform(-3, 4)$$

then

$$f(x) = \frac{1}{4 - (-3)} = \frac{1}{7}$$

which is illustrated in Fig. 6.9.

Table 6.1 Basic parameters for special continuous random variables

Normal random variable	Standard normal random variable
$X \sim N(\mu, \sigma^2)$	$Z \sim N(0, 1)$
$f(x) = \frac{1}{\sigma\sqrt{2\pi}} e^{-(x-\mu)^2/2\sigma^2} \quad -\infty < x < \infty$	$f(z) = \frac{1}{\sqrt{2\pi}} e^{-z^2/2} \quad -\infty < z < \infty$
$F(a) = \int_{-\infty}^{a} \frac{1}{\sigma\sqrt{2\pi}} e^{-(x-\mu)^2/2\sigma^2} dx$	$F(a) = \Phi(a) = \int_{-\infty}^{a} \frac{1}{\sqrt{2\pi}} e^{-z^2/2} dz$
$\overline{F}(a) = \int_{a}^{\infty} \frac{1}{\sigma\sqrt{2\pi}} e^{-(x-\mu)^2/2\sigma^2} dx$	$\overline{F}(a) = \overline{\Phi}(a) = \int_{a}^{\infty} \frac{1}{\sqrt{2\pi}} e^{-z^2/2} dz$
$E[X] = \mu$	$E[Z] = 0$
$Var(X) = \sigma^2$	$Var(Z) = 1$
	Lognormal random variable
	$Y \sim lognorm(\mu, \sigma^2)$ where
	$Y = e^X$
	$X \sim N(\mu, \sigma^2)$
	$f(y) = \frac{1}{y\sigma\sqrt{2\pi}} e^{-(lny-\mu)^2/2\sigma^2}$
	$F(a) = \Phi\left(\frac{\ln(a)-\mu}{\sigma}\right)$
	$\overline{F}(a) = \overline{\Phi}\left(\frac{\ln(a)-\mu}{\sigma}\right)$
	$E[Y] = e^{\mu + \frac{\sigma^2}{2}}$
	$Var(Y) = e^{2\mu + \sigma^2}(e^{\sigma^2} - 1)$
Exponential random variable	Gamma random variable
$X \sim \exp(\lambda)$	$t_n \sim Gamma(n, \lambda)$
$f(x) = \lambda e^{-\lambda x} \quad x \geq 0$	$f_{t_n}(t) = \frac{\lambda e^{-\lambda t}(\lambda t)^{n-1}}{(n-1)!}, t \geq 0$
$F(a) = 1 - e^{-\lambda a}$	$F_{t_n}(a) = \int_{0}^{a} \frac{\lambda e^{-\lambda t}(\lambda t)^{n-1}}{(n-1)!} dt$
$\overline{F}(a) = e^{-\lambda a}$	$\overline{F}_{t_n}(a) = \int\limits_{a}^{\infty} \frac{\lambda e^{-\lambda t}(\lambda t)^{n-1}}{(n-1)!} dt$
$E[X] = \frac{1}{\lambda}$	$E[t_n] = \frac{n}{\lambda}$
$Var(X) \frac{1}{\lambda^2}$	$Var(t_n) = \frac{n}{\lambda^2}$
Weibull random variable	Uniform random variable
$X \sim Weib(\alpha, \lambda)$	$X \sim Uniform(\alpha, \beta)$
$f(x) = \alpha\lambda^\alpha x^{\alpha-1} e^{-(\lambda x)^\alpha} \quad x \geq 0$	$f(x) = \frac{1}{\beta-\alpha} \quad \alpha < x < \beta$
$F(a) = 1 - e^{-(\lambda a)^\alpha} \quad x \geq 0$	$F(a) = \frac{a-\alpha}{\beta-\alpha} \quad \alpha < x < \beta$
$\overline{F}(a) = e^{-(\lambda a)^\alpha} \quad x \geq 0$	$\overline{F}(a) = \frac{\beta-a}{\beta-\alpha} \quad \alpha < x < \beta$
$E[X] = \frac{1}{\lambda}\Gamma\left(\frac{1}{\alpha} + 1\right)$	$E[X] = \frac{\alpha+\beta}{2}$
$Var(X) = \frac{1}{\lambda^2}\left[\Gamma\left(\frac{2}{\alpha} + 1\right) - \Gamma^2\left(\frac{1}{\alpha} + 1\right)\right]$	$Var(X) = \frac{(\beta-\alpha)^2}{12}$
where	
Γ denotes gamma function defined as	
$\Gamma(n) = \int_{0}^{\infty} e^{-x} x^{n-1} dx$	

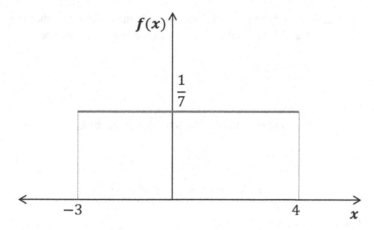

Fig. 6.9 Probability density function of a uniform random variable for problem 1

The probability that the position of the particle will be nonnegative will be

$$P(X \geq 0) = \int_0^4 \frac{1}{7} dx = \frac{4}{7}$$

the expected value of the position of the particle will be

$$E[X] = \frac{-3+4}{2} = \frac{1}{2}$$

and the variance of the position of the particle will be

$$Var(X) = \frac{(4-(-3))^2}{12} = \frac{49}{12} = 4.0833$$

To determine the probability density function of $|X|$, we start with the cumulative distribution function of $|X|$ such that

$$F_{|X|}(a) = P(|X| \leq a) = P(-a \leq X \leq a) = \frac{2a}{7} \quad \text{for} \ -3 < a < 4$$

Hence, the probability density function of $|X|$ will be

$$f_{|X|}(a) = \frac{dF_{|X|}(a)}{da} = \frac{2}{7} \quad \text{for} \ -3 < a < 4$$

2. The repair time of a broken-down machine is assumed to be *exponentially* distributed with mean 75 min. What is the probability that the repair time will exceed

2 h given that it already exceeds 1.5 h? What is the probability that the repair time is between 1 and 3 h? Find also the variance of the repair time.

Solution
Let

$$X : \text{Repair time of a broken}-\text{down machine}$$

Since

$$E[X] = 75 \min = 1.25 \, h$$

it follows that

$$\lambda = \frac{1}{E[X]} = \frac{1}{1.25} = 0.8 \, \text{per hour}$$

and

$$f(x) = 0.8 e^{-0.8x}$$

The probability that the repair time of this broken-down machine will exceed 2 h given that it already exceeds 1.5 h will be

$$P(X > 2 | X > 1.5) = P(X > 0.5)$$

by *memoryless property of exponential* random variable. Finally,

$$P(X > 0.5) = \overline{F}(0.5) = e^{-(0.8)(0.5)} = e^{-0.4} = 0.6703$$

The probability that the repair time is between 1 and 3 h will be

$$P(1 < X < 3) = \int_{1}^{3} 0.8 e^{-0.8x} dx = F(3) - F(1)$$

$$= \left(1 - e^{-(0.8)(3)}\right) - \left(1 - e^{-(0.8)(1)}\right) = e^{-0.8} - e^{-2.4} = 0.3586$$

and the variance of the repair time will be

$$Var(X) = \frac{1}{\lambda^2} = \frac{1}{(0.8)^2} = 1.5625$$

Warning *Time unit consistency* is especially important in the problems related to *exponential random variables*. One time unit should be selected, and the computations should be made accordingly. As an example, *hour* is selected as the time unit in this question.

3. Let $X \sim N(10, 25)$ be a normal random variable. Find the probability that X will be greater than 8 *or* less than 3. Find also $P(|X - 2| \le 3)$.

Solution

The probability that X will be greater than 8 *or* less than 3 will be

$$P(X > 8) + P(X < 3)$$

by Axiom 3. Accordingly,

$$P\left(\frac{X - 10}{\sqrt{25}} > \frac{8 - 10}{\sqrt{25}}\right) + P\left(\frac{X - 10}{\sqrt{25}} < \frac{3 - 10}{\sqrt{25}}\right)$$
$$= P(Z > -0.4) + P(Z < -1.4)$$
$$= (1 - \Phi(-0.4)) + \Phi(-1.4)$$
$$= (1 - (1 - \Phi(0.4))) + 1 - \Phi(1.4)$$
$$= \Phi(0.4) + 1 - \Phi(1.4)$$
$$= 0.6554 + 1 - 0.9192 = 0.7362$$

$$P(|X - 2| \le 3) = P(-3 \le X - 2 \le 3)$$
$$= P\left(\frac{-3 - 8}{\sqrt{25}} \le \frac{X - 2 - 8}{\sqrt{25}} \le \frac{3 - 8}{\sqrt{25}}\right)$$
$$= P\left(\frac{-11}{5} \le \frac{X - 10}{5} \le \frac{-5}{5}\right)$$
$$= P(-2.2 \le Z \le -1)$$
$$= \Phi(-1) - \Phi(-2.2) = (1 - \Phi(1)) - (1 - \Phi(2.2))$$
$$= \Phi(2.2) - \Phi(1) = 0.9861 - 0.8413 = 0.1448$$

4. The time between two successive calls arriving to a call center is an *independent exponential* random variable with rate 6 per hour. If the call center starts to answer calls at 9:00 a.m., what is the expected time of the 12th call? Given that the time between the 2nd and the 3rd calls exceeds 4 min, what is the probability that the time between the 5th and the 6th calls will exceed 7 min?

Solution

Let

t_n : Time of the nth call assuming that the beginning time is 0

Since the time between two successive calls is an *independent exponential* random variable,

$$t_n \sim Gamma(n, \lambda)$$

where $\lambda = 6$ per hour. Since

$$E[t_n] = \frac{n}{\lambda}$$

it follows that

$$E[t_{12}] = \frac{12}{6} = 2\,h$$

Recall that the call center starts its service at 9:00 a.m., then the expected time of the 12th call will be 11:00 a.m.

Let

$$X_3 = t_3 - t_2$$

$$X_6 = t_6 - t_5$$

be the time between the 2nd and the 3rd, and the time between the 5th and the 6th calls, respectively. Since the time between two successive calls is *independent*,

$$P\left(\left\{X_6 > \frac{7}{60}\right\} \middle| \left\{X_3 > \frac{4}{60}\right\}\right) = P\left(X_6 > \frac{7}{60}\right)$$

$$= \bar{F}\left(\frac{7}{60}\right) = e^{-(6)(\frac{7}{60})} = e^{-0.7} = 0.4966$$

5. You enter a bank with one server, and you see that a customer is already being served. The service times are *independent* and *exponentially* distributed with the rate 3 per hour.

    (a) Find the probability that you wait at least 5 min given that the customer is already being served for 10 min.
    (b) Find the probability that you wait at least 5 min given that the customer is already being served for 20 min.
    (c) Find the expected value and the variance of your service time given that the total service time of the customer who is already being served is 17 min.

Solution

Let

X : Total service time of the customer who is already being served

Hence,

$$X \sim \exp(0.05)$$

where $\lambda = \frac{3}{60} = 0.05$ per minute.

(a) The probability that you wait at least 5 min given that the customer is already being served for 10 min will be

$$P(X > 15|X > 10) = P(X > 5)$$

by *memoryless property of exponential* random variable. Finally,

$$P(X > 5) = \overline{F}(5) = e^{-(0.05)5} = e^{-0.25} = 0.7788$$

(b) The probability that you wait at least 5 min given that the customer is already being served for 20 min will be

$$P(X > 25|X > 20) = P(X > 5)$$

by *memoryless property of exponential* random variable. Finally,

$$P(X > 5) = \overline{F}(5) = e^{-(0.05)5} = e^{-0.25} = 0.7788$$

Note that the results of (a) and (b) do not change due to the *memoryless property of exponential* random variable.

(c) Let

Y : Your total service time

Since the service times are *independent*, the calculation of the expected value and the variance of your service time will be

$$E[Y] = \frac{1}{\lambda} = \frac{1}{0.05} = 20\,\text{min}$$

$$Var(Y) = \frac{1}{\lambda^2} = \frac{1}{(0.05)^2} = 400$$

without considering the total service time of the customer who is already being served.

6. A sample of 1,000 products is considered for quality control. Each product has at least one defect with probability 0.12. *Approximate* the probability that at least 120 products will be detected with at least one defect. Given that at least 50 products are detected with at least one defect, what is the *approximated* probability that the number of products with at least one defect is no more than 150?

Solution
Let

$$X : \text{Number of products with at least one defect}$$

Hence,

$$X \sim Bin(1,000, 0.12)$$

with

$$E[X] = 1,000(0.12) = 120$$

$$Var(X) = 1,000(0.12)(1 - 0.12) = 105.6$$

By standard normal random variable *approximation* to a binomial random variable, the *approximated* probability that *at least* 120 products will be detected with at least one defect will be

$$P(X \geq 119.5) = P\left(\frac{X - 120}{\sqrt{105.6}} \geq \frac{119.5 - 120}{\sqrt{105.6}}\right)$$
$$= P(Z \geq -0.0487) = P(Z \leq 0.0487)$$
$$= \Phi(0.0487) \cong 0.5199$$

The *approximated* probability that the number of products with at least one defect is *no more* than 150 given that *at least* 50 products are detected with at least one defect will be

$$P(\{X \leq 150.5\}|\{X \geq 49.5\}) = \frac{P(\{X \leq 150.5\} \cap \{X \geq 49.5\})}{P(X \geq 49.5)}$$

$$= \frac{P(49.5 \le X \le 150.5)}{P(X \ge 49.5)}$$

$$= \frac{P\left(\frac{49.5-120}{\sqrt{105.6}} \le Z \le \frac{150.5-120}{\sqrt{105.6}}\right)}{P\left(Z \ge \frac{49.5-120}{\sqrt{105.6}}\right)}$$

$$= \frac{P(-6.8605 \le Z \le 2.9680)}{P(Z \ge -6.8605)}$$

$$= \frac{\Phi(2.9680) - \Phi(-6.8605)}{1 - \Phi(-6.8605)}$$

$$= \frac{\Phi(2.9680) - (1 - \Phi(6.8605))}{1 - (1 - \Phi(6.8605))}$$

$$= \frac{\Phi(2.9680) + \Phi(6.8605) - 1}{\Phi(6.8605))}$$

$$\cong \frac{0.9985 + 1 - 1}{1} \cong 0.9985$$

7. Let X be a *Weibull* random variable that denotes the amount of emissions from a randomly selected engine. Assume that $\alpha = 1$ and $\lambda = \frac{1}{10}$. What is the probability that X is greater than 20 given that it is greater than 15?

Solution
Since

$$X \sim Weib\left(1, \frac{1}{10}\right)$$

it is equivalent to

$$X \sim \exp\left(\frac{1}{10}\right)$$

and the probability that X is greater than 20 given that it is greater than 15 will be

$$P(X > 20 | X > 15) = P(X > 5)$$

by *memoryless property of exponential* random variable. Finally,

$$P(X > 5) = \overline{F}(5) = e^{-\left(\frac{1}{10}\right)5} = e^{-0.5} = 0.6065$$

8. Assume $X \sim N(7, \sigma^2)$ and $P(X > 12) = 0.4$. Find $Var(X)$.

Solution

$$P(X > 12) = P\left(\frac{X - 7}{\sigma} > \frac{12 - 7}{\sigma}\right) = 0.4$$

$$P\left(Z > \frac{5}{\sigma}\right) = 0.4$$

$$1 - \Phi\left(\frac{5}{\sigma}\right) = 0.4$$

$$\Phi\left(\frac{5}{\sigma}\right) = 0.6$$

By reading the values from the cumulative distribution function table of a standard normal random variable, we find

$$\frac{5}{\sigma} \cong 0.25$$

and

$$\sigma \cong 20$$

which means

$$Var(X) = \sigma^2 \cong 400$$

9. A fair die is rolled 1,000 times. Suppose that your aim is to obtain at least 5 as an outcome for each roll. *Approximate* the probability that you reach your aim 320 times.

Solution
Let

$$X : \text{Number of outcomes with at least 5}$$

By noting that the probability of obtaining at least 5 is $\frac{2}{6} = \frac{1}{3}$ for each roll,

$$X \sim Bin\left(1,000, \frac{1}{3}\right)$$

Note that

$$E[X] = 1,000\left(\frac{1}{3}\right) = 333.33$$

$$Var(X) = 1,000\left(\frac{1}{3}\right)\left(1 - \frac{1}{3}\right) = \frac{2,000}{9} = 222.22$$

By standard normal random variable *approximation* to a binomial random variable,

$$
\begin{aligned}
P(X = 320) &\cong P(319.5 \leq X \leq 320.5) \\
&= P\left(\frac{319.5 - 333.33}{\sqrt{222.22}} \leq Z \leq \frac{320.5 - 333.33}{\sqrt{222.22}}\right) \\
&= P(-0.9278 \leq Z \leq -0.8607) \\
&= \Phi(-0.8607) - \Phi(-0.9278) \\
&= (1 - \Phi(0.8607)) - (1 - \Phi(0.9278)) \\
&= \Phi(0.9278) - \Phi(0.8607) \cong 0.8238 - 0.8051 = 0.0187
\end{aligned}
$$

Chapter 7
Other Selected Topics in Basic Probability

Abstract In the first part of this chapter, the basic terms related to the jointly distributed random variables including the joint probability mass function and joint probability density function, sums of independent random variables, convolution of random variables, order statistics of continuous random variables, covariance and correlation of random variables, and expected value and variance of sum of random variables have been explained. The second part of this chapter has been devoted to the conditional distribution, conditional expected value, conditional variance, expected value by conditioning, and variance by conditioning. In the third part of the chapter, the moment generating function and characteristic function have been defined, and a list of moment generating functions has been provided for special random variables. The last part of the chapter included the limit theorems in probability including Strong Law of Large Numbers and Central Limit Theorem. The concepts have been illustrated with several examples and problems.

This is a chapter that combines the learnings from all former chapters related to the basics of probability by considering one random variable or two or more discrete or continuous random variables. Several topics in this chapter such as jointly distributed random variables, order statistics, covariance, correlation, expected value by conditioning, moment generating function, and limit theorems including Central Limit Theorem and Strong Law of Large Numbers are extremely important not only in stochastic processes, but also in statistics. Thus, this chapter can be interpreted as a bridge between the chapters related to the basics of probability, and the chapters related to the basics of stochastic processes in this book. Several examples and problems will be presented for both discrete and continuous random variables that will be background for the chapters related to the basics of stochastic processes.

© Springer Nature Switzerland AG 2019 95
E. Bas, *Basics of Probability and Stochastic Processes*,
https://doi.org/10.1007/978-3-030-32323-3_7

7.1 Jointly Distributed Random Variables

Example 1 Suppose that 3 balls are randomly selected from an urn that contains 2 green, 4 white, and 3 black balls. The random variables X, Y, Z can be defined as the number of the *green*, *white* and *black* balls withdrawn, respectively.

Definition 1 (*Jointly distributed random variables*) Jointly distributed random variables are random variables with two or more random variables.

Table 7.1 provides the basic terms for the jointly distributed random variables with two random variables.

Example 1 (revisited 1) Recall that 3 balls are randomly selected from an urn that contains 2 green, 4 white, and 3 black balls, and

X: Number of the *green* balls withdrawn
Y: Number of the *white* balls withdrawn
Z: Number of the *black* balls withdrawn

We consider Table 7.2 for the possible outcomes.

As an example, the joint probability mass function of X and Y will be

Table 7.1 Basic terms for the jointly distributed random variables with two random variables

Jointly distributed discrete random variables: Basic terms	*Jointly distributed continuous random variables: Basic terms*
If X and Y are two *discrete* random variables, then $p(x, y) = P(\{X = x\} \cap \{Y = y\})$ is the *joint probability mass function* of the random variables X and Y, where $\sum_x \sum_y p(x, y) = 1$	If X and Y are two *continuous* random variables, then $f(x, y)$ is the *joint probability density function* of the random variables X and Y, where $\int_{y_{min}}^{y_{max}} \int_{x_{min}}^{x_{max}} f(x, y) dx dy = 1$
$F(a, b) = P(\{X \leq a\} \cap \{Y \leq b\})$ is the *joint cumulative distribution function* of the discrete random variables X and Y.	$F(a, b) = \int_{y_{min}}^{b} \int_{x_{min}}^{a} f(x, y) dx dy$ is the *joint cumulative distribution function* of the continuous random variables X and Y, where $f(a, b) = \frac{\partial^2}{\partial a \partial b} F(a, b)$
The probability mass function of the discrete random variable X is $p_X(x) = \sum_y p(x, y)$ The probability mass function of the discrete random variable Y is $p_Y(y) = \sum_x p(x, y)$ If X and Y are two *independent discrete* random variables, then $p(x, y) = p_X(x) p_Y(y)$ holds for all x, y.	The probability density function of the continuous random variable X is $f_X(x) = \int_{y_{min}}^{y_{max}} f(x, y) dy$ The probability density function of the continuous random variable Y is $f_Y(y) = \int_{x_{min}}^{x_{max}} f(x, y) dx$ If X and Y are two *independent continuous* random variables, then $f(x, y) = f_X(x) f_Y(y)$ holds for all x, y.

Table 7.2 Possible outcomes for the ball withdrawal problem

X	Y	Z
0	0	3
0	1	2
0	2	1
0	3	0
1	0	2
1	1	1
1	2	0
2	0	1
2	1	0

$$p(0, 0) = P(\{X = 0\} \cap \{Y = 0\}) = \binom{2}{0}\binom{4}{0}\binom{3}{3} / \binom{9}{3} = \frac{1}{84}$$

$$p(0, 1) = P(\{X = 0\} \cap \{Y = 1\}) = \binom{2}{0}\binom{4}{1}\binom{3}{2} / \binom{9}{3} = \frac{12}{84}$$

$$p(0, 2) = P(\{X = 0\} \cap \{Y = 2\}) = \binom{2}{0}\binom{4}{2}\binom{3}{1} / \binom{9}{3} = \frac{18}{84}$$

$$p(0, 3) = P(\{X = 0\} \cap \{Y = 3\}) = \binom{2}{0}\binom{4}{3}\binom{3}{0} / \binom{9}{3} = \frac{4}{84}$$

$$p(1, 0) = P(\{X = 1\} \cap \{Y = 0\}) = \binom{2}{1}\binom{4}{0}\binom{3}{2} / \binom{9}{3} = \frac{6}{84}$$

$$p(1, 1) = P(\{X = 1\} \cap \{Y = 1\}) = \binom{2}{1}\binom{4}{1}\binom{3}{1} / \binom{9}{3} = \frac{24}{84}$$

$$p(1, 2) = P(\{X = 1\} \cap \{Y = 2\}) = \binom{2}{1}\binom{4}{2}\binom{3}{0} / \binom{9}{3} = \frac{12}{84}$$

$$p(2, 0) = P(\{X = 2\} \cap \{Y = 0\}) = \binom{2}{2}\binom{4}{0}\binom{3}{1} / \binom{9}{3} = \frac{3}{84}$$

$$p(2, 1) = P(\{X = 2\} \cap \{Y = 1\}) = \binom{2}{2}\binom{4}{1}\binom{3}{0} / \binom{9}{3} = \frac{4}{84}$$

The probability mass function of X will be

$$p_X(0) = p(0, 0) + p(0, 1) + p(0, 2) + p(0, 3) = \frac{35}{84}$$

$$p_X(1) = p(1, 0) + p(1, 1) + p(1, 2) = \frac{42}{84}$$

$$p_X(2) = p(2, 0) + p(2, 1) = \frac{7}{84}$$

The probability mass function of Y will be

$$p_Y(0) = p(0,0) + p(1,0) + p(2,0) = \frac{10}{84}$$
$$p_Y(1) = p(0,1) + p(1,1) + p(2,1) = \frac{40}{84}$$
$$p_Y(2) = p(0,2) + p(1,2) = \frac{30}{84}$$
$$p_Y(3) = p(0,3) = \frac{4}{84}$$

Example 2 The joint probability density function of the continuous random variables X and Y is given as

$$f(x,y) = x + y, \quad 0 < x < 1, 0 < y < 1$$

As an example,

$$P\left(\left\{X > \frac{1}{8}\right\} \cap \left\{Y < \frac{1}{4}\right\}\right) = \int_0^{1/4}\int_{1/8}^1 (x+y)dxdy = \frac{77}{512} = 0.1504$$

$$P(Y < X) = \int_0^1\int_y^1 (x+y)dxdy = \frac{1}{2}$$

The probability density function of X will be

$$f_X(x) = \int_{y_{min}}^{y_{max}} f(x,y)dy = \int_0^1 (x+y)dy = x + \frac{1}{2}$$

The probability density function of Y will be

$$f_Y(y) = \int_{x_{min}}^{x_{max}} f(x,y)dx = \int_0^1 (x+y)dx = y + \frac{1}{2}$$

Remark 1 Note that since $f_X(x)f_Y(y) = \left(x + \frac{1}{2}\right)\left(y + \frac{1}{2}\right) \neq f(x,y)$, X and Y are *dependent*.

Definition 2 (*Sums of independent random variables*) The sums of some special independent random variables defined in Chaps. 5 and 6 lead to the same or other special random variables. Table 7.3 provides a list of these random variables.

Table 7.3 Sums of independent random variables

Sum of independent binomial random variables	Sum of independent Poisson random variables
If $X_1 \sim Bin(n_1, p)$ $X_2 \sim Bin(n_2, p)$ $\ldots$ $X_m \sim Bin(n_m, p)$ are *independent* random variables, then $\left(\sum_{i=1}^{m} X_i\right) \sim Bin\left(\sum_{i=1}^{m} n_i, p\right)$	If $X_1 \sim Poiss(\lambda_1)$ $X_2 \sim Poiss(\lambda_2)$ $\ldots$ $X_m \sim Poiss(\lambda_m)$ are *independent* random variables, then $\left(\sum_{i=1}^{m} X_i\right) \sim Poiss\left(\sum_{i=1}^{m} \lambda_i\right)$
Sum of independent uniform random variables	*Sum of independent exponential random variables*
If $X_1 \sim Uniform(0, 1)$ $X_2 \sim Uniform(0, 1)$ $\ldots$ $X_m \sim Uniform(0, 1)$ are *independent* random variables, then $\left(\sum_{i=1}^{m} X_i\right) \sim Irwin - Hall$ with $f_{X_1+X_2+\cdots+X_m}(x)$ $= \sum_{i=0}^{m}\left[(-1)^i \binom{m}{i} \dfrac{(x-i)^{m-1}}{(m-1)!} s_i(x)\right]$ $F_{X_1+X_2+\cdots+X_m}(x)$ $= \sum_{i=0}^{m}\left[(-1)^i \binom{m}{i} \dfrac{(x-i)^{m}}{m!} s_i(x)\right]$ where $s_i(x) = \begin{cases} 0 \ if \ i > x \\ 1 \ if \ i \leq x \end{cases}$ is the unit step function.	If $X_1 \sim \exp(\lambda)$ $X_2 \sim \exp(\lambda)$ $\ldots$ $X_m \sim \exp(\lambda)$ are *independent* random variables, then $\left(\sum_{i=1}^{m} X_i\right) \sim Gamma(m, \lambda)$
Special case If $X_1 \sim Uniform(0, 1)$ $X_2 \sim Uniform(0, 1)$ are *independent* random variables, then $X_1 + X_2$ is a *triangular* random variable with $f_{X_1+X_2}(x) = x \quad for \ 0 \leq x \leq 1$ $\qquad\qquad 2 - x \quad for \ 1 < x < 2$	If $X_1 \sim N\left(\mu_1, \sigma_1^2\right)$ $X_2 \sim N\left(\mu_2, \sigma_2^2\right)$ $\ldots$ $X_m \sim N\left(\mu_m, \sigma_m^2\right)$ are *independent* random variables, then $\left(\sum_{i=1}^{m} X_i\right) \sim N\left(\sum_{i=1}^{m} \mu_i, \sum_{i=1}^{m} \sigma_i^2\right)$

Example 3 Three *fair coins* are tossed *independently* from each other. The first coin is tossed 5 times, the second coin is tossed 7 times, and the third coin is tossed 10 times. Each toss of each coin is also *independent* from each other. What is the probability that the total number of heads will be 8?

Let
X_i: The number of heads as a result of tossing the ith coin, $i = 1, 2, 3$.
Accordingly,

$$X_1 \sim Bin\left(5, \frac{1}{2}\right)$$

$$X_2 \sim Bin\left(7, \frac{1}{2}\right)$$

$$X_3 \sim Bin\left(10, \frac{1}{2}\right)$$

and

$$(X_1 + X_2 + X_3) \sim Bin\left(22, \frac{1}{2}\right)$$

follows by the *sum* of *independent binomial* random variables. Finally, the probability that the total number of heads will be 8 is

$$P(X_1 + X_2 + X_3 = 8) = \binom{22}{8}\left(\frac{1}{2}\right)^8\left(1 - \frac{1}{2}\right)^{22-8}$$

$$= \binom{22}{8}\left(\frac{1}{2}\right)^8\left(\frac{1}{2}\right)^{14} = 0.0762$$

Definition 3 (*Convolution of random variables*) Convolution of random variables refers to the probability mass (density) function or cumulative distribution function of the sum of two or more independent random variables.

Table 7.4 provides the basic formulae for the convolution of two random variables.

Definition 4 (*Order statistics of continuous random variables*) Let

$$X_1, X_2, \ldots, X_n$$

be *independent and identically distributed* (*i.i.d.*) continuous random variables.

$$X_{(1)}, X_{(2)}, \ldots, X_{(n)}$$

is called as the *order statistics* corresponding to the continuous random variables $X_1, X_2, \ldots, X_n$, where

Table 7.4 Convolution of two random variables

Convolution of two discrete random variables	Convolution of two continuous random variables
$F_{X+Y}(a) = \sum_y F_X(a-y) p_Y(y)$ is called as the *convolution* of F_X and F_Y for the *discrete* and *independent* random variables X and Y.	$F_{X+Y}(a) = \int_{y_{min}}^{y_{max}} F_X(a-y) f_Y(y) dy$ is called as the *convolution* of F_X and F_Y for the *continuous* and *independent* random variables X and Y.
$p_{X+Y}(a) = \sum_y p_X(a-y) p_Y(y)$ is called as the *convolution* of p_X and p_Y for the *discrete* and *independent* random variables X and Y.	$f_{X+Y}(a) = \int_{y_{min}}^{y_{max}} f_X(a-y) f_Y(y) dy$ is called as the *convolution* of f_X and f_Y for the *continuous* and *independent* random variables X and Y.

$$X_{(1)} \le X_{(2)} \le \cdots \le X_{(n)}$$

$X_{(i)}$: *i*th smallest of $X_1, X_2, \ldots, X_n$

Note that the joint probability density function of the order statistics $X_{(1)}, X_{(2)}, \ldots, X_{(n)}$ will be

$$f_{X_{(1)}, X_{(2)}, \ldots, X_{(n)}}(x_1, x_2, \ldots, x_n)$$
$$= n! \prod_{i=1}^{n} f(x_i) \quad \text{for} \quad x_1 < x_2 < \cdots < x_n$$

Remark 2 Note that the concept of order statistics is *not defined* for the *discrete* random variables since ties are possible for discrete random variables.

Example 4 Let $X_i \sim Uniform(\alpha, \beta)$ be *independent* random variables for $i = 1, 2, \ldots, n$. The joint probability density function of the order statistics of $X_1, X_2, \ldots, X_n$ will be

$$f_{X_{(1)}, X_{(2)}, \ldots, X_{(n)}}(x_1, x_2, \ldots, x_n) = n! \left(\frac{1}{\beta - \alpha}\right)^n$$

where

$$\alpha < x_1 < x_2 < \cdots < x_n < \beta$$

Note that

$$n!$$

represents the number of different orders (permutations) of $X_1, X_2, \ldots, X_n$, and

$$\left(\frac{1}{\beta - \alpha}\right)^n$$

represents the joint probability density function for each order (permutation) of $X_1, X_2, \ldots, X_n$.

Definition 5 (*Covariance and correlation of random variables*) *Covariance* is a quantitative measure that describes the joint deviation of two random variables from their expected values. *Correlation* is a measure that describes the covariance of two random variables normalized by the standard deviation of each random variable.

Table 7.5 provides the basic formulae for the covariance and correlation of the random variables.

Example 5 We consider two *independent* rolls of a die. Let X_1 be the outcome of the first die, and let X_2 be the outcome of the second die. Accordingly,

$$
\begin{aligned}
Cov(X_1 + X_2, X_1 - X_2) &= Cov(X_1, X_1) - Cov(X_1, X_2) \\
&\quad + Cov(X_2, X_1) - Cov(X_2, X_2) \\
&= Var(X_1) - Var(X_2) = 0
\end{aligned}
$$

Remark 3 Note that $Cov(X_1, X_2) = Cov(X_2, X_1)$. Note also that $Cov(X_1, X_1) = Var(X_1) = Cov(X_2, X_2) = Var(X_2)$ holds due to the *independent* rolls of the die.

Definition 6 (*Expected value of the sum of random variables*) Let $X_1, X_2, \ldots, X_n$ be *independent* or *dependent* discrete or continuous random variables. The expected value of the sum of these random variables will be

Table 7.5 Covariance and correlation of the random variables

Covariance and correlation of the discrete and continuous random variables
$Cov(X, Y) = E[(X - E[X])(Y - E[Y])]$
$\quad = E[XY] - E[X]E[Y]$
is the covariance between the *discrete* or *continuous* random variables X and Y with the following properties:
$Cov(X, X) = Var(X)$
$Cov(X, Y) = Cov(Y, X)$
$Cov(aX, bY) = abCov(X, Y)$
$Cov\left(\sum_{i=1}^{n} X_i, \sum_{j=1}^{m} Y_j\right) = \sum_{i=1}^{n} \sum_{j=1}^{m} Cov(X_i, Y_j)$
Note that
$Cov(X, Y) = 0$
if X and Y are *independent discrete* or *continuous* random variables.
$\rho(X, Y) = Corr(X, Y) = \frac{Cov(X,Y)}{\sqrt{Var(X)Var(Y)}}$
is called as the *correlation coefficient* for the *discrete* or *continuous* random variables X and Y, where $Var(X) > 0$, $Var(Y) > 0$, and
$-1 \le \rho(X, Y) \le 1$

$$E\left(\sum_{i=1}^{n} X_i\right) = \sum_{i=1}^{n} E[X_i]$$

Special case

$$E\left(\left(\sum_{i=1}^{n} a_i X_i\right) + b\right) = \left(\sum_{i=1}^{n} a_i E[X_i]\right) + b$$

Definition 7 (*Variance of the sum of random variables*) Let $X_1, X_2, \ldots, X_n$ be *independent* discrete or continuous random variables. The variance of the sum of these random variables will be

$$Var\left(\sum_{i=1}^{n} X_i\right) = \sum_{i=1}^{n} Var(X_i)$$

Special case

$$Var\left(\left(\sum_{i=1}^{n} a_i X_i\right) + b\right) = \sum_{i=1}^{n} a_i^2 Var(X_i)$$

Let $X_1, X_2, \ldots, X_n$ be discrete or continuous random variables with at least two *dependent* random variables. Hence,

$$Var\left(\sum_{i=1}^{n} X_i\right) = \sum_{i=1}^{n} Var(X_i) + 2 \sum_{i,i<j} \sum_{j} Cov(X_i, X_j)$$

Example 1 (revisited 2) Recall that 3 balls are randomly selected from an urn that contains 2 green, 4 white, and 3 black balls. Recall also that

X: Number of the *green* balls withdrawn
Y: Number of the *white* balls withdrawn

The expected value of the total number of the *green* balls and the *white* balls withdrawn will be

$$E[X + Y] = E[X] + E[Y]$$

By using the probability values computed in Example 1 (revisited 1),

$$E[X] = (0)p_X(0) + (1)p_X(1) + (2)p_X(2)$$
$$= p_X(1) + 2p_X(2) = (p(1, 0) + p(1, 1) + p(1, 2)) + 2(p(2, 0) + p(2, 1))$$
$$= \frac{42}{84} + (2)\frac{7}{84} = \frac{56}{84}$$
$$E[Y] = (0)p_Y(0) + (1)p_Y(1) + (2)p_Y(2) + (3)p_Y(3) = p_Y(1) + 2p_Y(2) + 3p_Y(3)$$

$$= (p(0, 1) + p(1, 1) + p(2, 1)) + 2(p(0, 2) + p(1, 2)) + 3(p(0, 3))$$
$$= \frac{40}{84} + (2)\frac{30}{84} + (3)\frac{4}{84} = \frac{112}{84}$$

Finally,

$$E[X + Y] = E[X] + E[Y] = \frac{56}{84} + \frac{112}{84} = \frac{168}{84} = 2$$

The variance of the total number of the *green* balls and the *white* balls withdrawn will be

$$Var(X + Y) = Var(X) + Var(Y) + 2Cov(X, Y)$$

where

$$E[X^2] = (1^2)p_X(1) + (2^2)p_X(2) = \frac{42}{84} + (4)\frac{7}{84} = \frac{70}{84}$$

$$Var(X) = E[X^2] - (E[X])^2 = \frac{70}{84} - \left(\frac{56}{84}\right)^2 = 0.3889$$

$$E[Y^2] = (1^2)p_Y(1) + (2^2)p_Y(2) + (3^2)p_Y(3) = \frac{40}{84} + (4)\frac{30}{84} + (9)\frac{4}{84} = \frac{196}{84}$$

$$Var(Y) = E[Y^2] - (E[Y])^2 = \frac{196}{84} - \left(\frac{112}{84}\right)^2 = 0.5556$$

$$Cov(X, Y) = E[XY] - E[X]E[Y]$$

where

$$E[XY] = (0)\left(\frac{1}{84} + \frac{12}{84} + \frac{18}{84} + \frac{4}{84} + \frac{6}{84} + \frac{3}{84}\right)$$
$$+ 1\left(\frac{24}{84}\right) + 2\left(\frac{12}{84} + \frac{4}{84}\right) = \frac{56}{84}$$

Finally,

$$Cov(X, Y) = E[XY] - E[X]E[Y]$$
$$= \frac{56}{84} - \left(\frac{56}{84}\right)\left(\frac{112}{84}\right) = -0.2222$$

and

$$Var(X + Y) = Var(X) + Var(Y) + 2Cov(X, Y)$$
$$= 0.3889 + 0.5556 + 2(-0.2222) = 0.5001$$

7.2 Conditional Distribution, Conditional Expected Value, Conditional Variance, Expected Value by Conditioning, Variance by Conditioning

Table 7.6 provides the basic formulae of the conditional distribution, conditional expected value, conditional variance, expected value by conditioning, and variance by conditioning for the discrete and continuous random variables.

Example 1 (revisited 3) Recall that 3 balls are randomly selected from an urn that contains 2 green, 4 white, and 3 black balls. Recall also that

X: Number of the *green* balls withdrawn
Y: Number of the *white* balls withdrawn

By using the probability values computed in Example 1 (revisited 1), given that 1 *white* ball is withdrawn, the expected number of the *green* balls withdrawn will be

$$E(X|Y = 1) = \sum_x x P(X = x|Y = 1)$$
$$= (0)P(X = 0|Y = 1) + (1)P(X = 1|Y = 1) + (2)P(X = 2|Y = 1)$$
$$= P(X = 1|Y = 1) + 2P(X = 2|Y = 1)$$

Table 7.6 Conditional distribution, conditional expected value, conditional variance, expected value by conditioning, variance by conditioning

Conditional probability mass function for the discrete random variables	Conditional probability density function for the continuous random variables
$p_{(X|Y)}(x|y) = P(X = x|Y = y) = \frac{p(x,y)}{p_Y(y)}$ for $p_Y(y) > 0$	$f_{(X|Y)}(x|y) = \frac{f(x,y)}{f_Y(y)}$ for $f_Y(y) > 0$
Conditional expected value and conditional variance for the discrete random variables	*Conditional expected value and conditional variance for the continuous random variables*
If X and Y are two *discrete* random variables, then the conditional expected value and the conditional variance of X given that $Y = y$ will be $E(X|Y = y) = \sum_x x p_{(X|Y)}(x|y)$	If X and Y are two *continuous* random variables, then the conditional expected value and the conditional variance of X given that $Y = y$ will be $E(X|Y = y) = \int_{x_{min}}^{x_{max}} x f_{(X|Y)}(x|y)dx$
$Var(X|Y = y) = E(X^2|Y = y) - (E(X|Y = y))^2$	$Var(X|Y = y) = E(X^2|Y = y) - (E(X|Y = y))^2$
Expected value and variance by conditioning for the discrete random variables	*Expected value and variance by conditioning for the continuous random variables*
$E[X] = \sum_y E(X|Y = y)p_Y(y)$ or $E[X] = E[E(X|Y)]$	$E[X] = \int_{y_{min}}^{y_{max}} E(X|Y = y) f_Y(y)dy$ or $E[X] = E[E(X|Y)]$
$Var(X) = E[Var(X|Y)] + Var(E(X|Y))$	$Var(X) = E[Var(X|Y)] + Var(E(X|Y))$

$$= \frac{P(\{X=1\} \cap \{Y=1\})}{P(Y=1)} + 2\left(\frac{P(\{X=2\} \cap \{Y=1\})}{P(Y=1)}\right)$$

$$= \frac{p(1,1)}{p(0,1)+p(1,1)+p(2,1)} + 2\left(\frac{p(2,1)}{p(0,1)+p(1,1)+p(2,1)}\right)$$

$$= \frac{\frac{24}{84}}{\left(\frac{12}{84}+\frac{24}{84}+\frac{4}{84}\right)} + 2\left(\frac{\frac{4}{84}}{\left(\frac{12}{84}+\frac{24}{84}+\frac{4}{84}\right)}\right) = \frac{24}{40} + \frac{8}{40} = \frac{32}{40} = 0.8$$

Given that 1 *white* ball is withdrawn, the variance of the number of the *green* balls withdrawn will be

$$Var(X|Y=1) = E\left(X^2|Y=1\right) - (E(X|Y=1))^2$$

where

$$E\left(X^2|Y=1\right) = \left(0^2\right)P(X=0|Y=1) + (1^2)P(X=1|Y=1) + (2^2)P(X=2|Y=1)$$

$$= P(X=1|Y=1) + 4P(X=2|Y=1)$$

$$= \frac{p(1,1)}{p(0,1)+p(1,1)+p(2,1)} + 4\left(\frac{p(2,1)}{p(0,1)+p(1,1)+p(2,1)}\right)$$

$$= \frac{\frac{24}{84}}{\left(\frac{12}{84}+\frac{24}{84}+\frac{4}{84}\right)} + 4\left(\frac{\frac{4}{84}}{\left(\frac{12}{84}+\frac{24}{84}+\frac{4}{84}\right)}\right) = \frac{24}{40} + \frac{16}{40} = 1$$

Finally,

$$Var(X|Y=1) = E\left(X^2|Y=1\right) - (E(X|Y=1))^2 = 1 - (0.8)^2 = 0.36$$

Example 2 (revisited 1) Recall that

$$f(x,y) = x+y, \quad 0 < x < 1, \quad 0 < y < 1$$

$$f_Y(y) = y + \frac{1}{2}, \quad 0 < y < 1$$

from Example 2. Assume that we wish to find $E\left(X|Y=\frac{1}{2}\right)$. Hence,

$$E(X|Y=y) = \int_0^1 x f_{(X|Y)}(x|y)dx$$

where

$$f_{(X|Y)}(x|y) = \frac{f(x,y)}{f_Y(y)} = \frac{x+y}{y+\frac{1}{2}} = \frac{2(x+y)}{2y+1}$$

and

$$E(X|Y = y) = \int\limits_{0}^{1} \frac{x2(x+y)}{2y+1} dx = \frac{1}{2y+1} \int\limits_{0}^{1} (2x^2 + 2xy)dx$$

$$= \frac{1}{2y+1}\left(\frac{2}{3} + y\right)$$

Finally,

$$E\left(X|Y = \frac{1}{2}\right) = \frac{1}{2(\frac{1}{2})+1}\left(\frac{2}{3} + \frac{1}{2}\right) = \frac{1}{2}\left(\frac{2}{3} + \frac{1}{2}\right) = \frac{7}{12}$$

Example 1 (revisited 4) Recall that 3 balls are randomly selected from an urn that contains 2 green, 4 white, and 3 black balls. Recall also that

X: Number of the *green* balls withdrawn
Y: Number of the *white* balls withdrawn

By using the probability values computed in Example 1 (revisited 1), the expected number of the *green* balls withdrawn by conditioning on the number of the *white* balls withdrawn will be

$$E[X] = \sum_{y} E(X|Y = y)p_Y(y)$$

$$= E(X|Y = 0)p_Y(0) + E(X|Y = 1)p_Y(1)$$
$$+ E(X|Y - 2)p_Y(2) + E(X|Y - 3)p_Y(3)$$
$$= [(0)p(0,0) + (1)p(1,0) + (2)p(2,0)]$$
$$+ [(0)p(0,1) + (1)p(1,1) + (2)p(2,1)]$$
$$+ [(0)p(0,2) + (1)p(1,2)] + [(0)p(0,3)]$$
$$= \left[(0)\frac{1}{84} + (1)\frac{6}{84} + (2)\frac{3}{84}\right]$$
$$+ \left[(0)\frac{12}{84} + (1)\frac{24}{84} + (2)\frac{4}{84}\right]$$
$$+ \left[(0)\frac{18}{84} + (1)\frac{12}{84}\right] + \left[(0)\frac{4}{84}\right]$$
$$= \frac{56}{84} = \frac{2}{3}$$

7.3   Moment Generating Function and Characteristic Function

Definition 8 (*Moment generating function*) $M_X(t)$ is the moment generating function of the *discrete* random variable X defined such that

$$M_X(t) = E[e^{tX}] = \sum_x e^{tx} p(x)$$

$M_X(t)$ is the moment generating function of the *continuous* random variable X defined such that

$$M_X(t) = E[e^{tX}] = \int_{x_{min}}^{x_{max}} e^{tx} f(x) dx$$

Remark 4 Note that for the *discrete* and *continuous* random variables, the following equations will hold:

$$M_X'(0) = E[X]$$
$$M_X''(0) = E[X^2]$$
$$\dots$$
$$M_X^{(n)}(0) = E[X^n], n \geq 1$$

Note also that if $X_1, X_2, \dots, X_n$ are *independent discrete* or *continuous* random variables with moment generating functions $M_{X_1}(t), M_{X_2}(t), \dots, M_{X_n}(t)$, respectively, then

$$M_{X_1+X_2+\dots+X_n}(t) = M_{X_1}(t) M_{X_2}(t) \dots M_{X_n}(t)$$

Definition 9 (*Characteristic function*) $C_X(t)$ is the characteristic function of the *discrete* random variable X defined such that

$$C_X(t) = E[e^{itX}] = \sum_x e^{itx} p(x)$$

$C_X(t)$ is the characteristic function of the *continuous* random variable X defined such that

$$C_X(t) = E[e^{itX}] = \int_{x_{min}}^{x_{max}} e^{itx} f(x) dx$$

Remark 5 Note that $C_X(t)$ is a complex-valued function, and i is the imaginary unit with the property $i^2 = -1$.

Table 7.7 provides the moment generating functions of special discrete and continuous random variables.

Example 6 Suppose that we toss two coins. The random variable X can be defined as the total number of heads. Recall that

$$P(X = 0) = p(0) = \frac{1}{4}$$

$$P(X = 1) = p(1) = \frac{1}{2}$$

$$P(X = 2) = p(2) = \frac{1}{4}$$

from Example 1 (revisited 3) in Chap. 4. The variance of X can be calculated by using the moment generating function such that

$$M_X(t) = E[e^{tX}] = \sum_x e^{tx} p(x) = e^{(t)(0)} p(0) + e^{(t)(1)} p(1) + e^{(t)(2)} p(2)$$

$$= p(0) + p(1)e^t + p(2)e^{2t} = \frac{1}{4} + \frac{1}{2}e^t + \frac{1}{4}e^{2t}$$

$$M'_X(t) = \frac{1}{2}e^t + \frac{2}{4}e^{2t}$$

$$M'_X(0) = \frac{1}{2} + \frac{2}{4} = 1 = E[X]$$

$$M''_X(t) = \frac{1}{2}e^t + \frac{4}{4}e^{2t}$$

$$M''_X(0) = \frac{1}{2} + \frac{4}{4} = \frac{3}{2} = E[X^2]$$

$$Var(X) = \sigma^2 = E[X^2] - (E[X])^2 = \frac{3}{2} - (1)^2 = \frac{1}{2}$$

Table 7.7 Moment generating functions of special discrete and continuous random variables

Moment generating functions of special discrete random variables	Moment generating functions of special continuous random variables
If $X \sim Bin(n, p)$	If $X \sim Uniform(\alpha, \beta)$
$M_X(t) = (pe^t + 1 - p)^n$	$M_X(t) = \frac{e^{t\beta} - e^{t\alpha}}{t(\beta - \alpha)}$
If $X \sim Poiss(\lambda)$	If $X \sim exp(\lambda)$
$M_X(t) = e^{\lambda(e^t - 1)}$	$M_X(t) = \frac{\lambda}{\lambda - t}, t < \lambda$
If $X \sim NegBin(r, p)$	If $X \sim N(\mu, \sigma^2)$
$M_X(t) = \left[\frac{pe^t}{1-(1-p)e^t}\right]^r, (1 - p)e^t < 1$	$M_X(t) = e^{\mu t + \frac{\sigma^2 t^2}{2}}$
If $X \sim Geom(p)$	If $Z \sim N(0, 1)$
$M_X(t) = \frac{pe^t}{1-(1-p)e^t}, (1 - p)e^t < 1$	$M_Z(t) = e^{t^2/2}$

Table 7.8 Basic limit theorems in probability

Markov's Inequality	Chebyshev's Inequality
If X is a random variable with $E[X] = \mu$ that takes *only nonnegative* values, then for any constant $k > 0$, $$P(X \geq k) \leq \tfrac{\mu}{k}$$	If X is a random variable with finite $E[X] = \mu$ and finite $Var(X) = \sigma^2$, then for any constant $k > 0$, $$P(|X - \mu| \geq k) \leq \tfrac{\sigma^2}{k^2}$$
Central Limit Theorem (CLT)	Strong Law of Large Numbers (SLLN)
If $X_1, X_2, \ldots, X_n$ are a *sequence* of i.i.d. random variables with $$E[X_1] = E[X_2] = \cdots = E[X_n] = \mu$$ $$Var(X_1) = Var(X_2) = \cdots = Var(X_n) = \sigma^2$$ Then $$\frac{X_1 + X_2 + \cdots + X_n - n\mu}{\sigma\sqrt{n}} \to Z \sim N(0, 1)$$ *as* $n \to \infty$	If $X_1, X_2, \ldots, X_n$ are a *sequence* of i.i.d. random variables with finite $$E[X_1] = E[X_2] = \cdots = E[X_n] = \mu$$ Then $$\frac{X_1 + X_2 + \cdots + X_n}{n} \to \mu \quad \text{as } n \to \infty$$
	Weak Law of Large Numbers (WLLN)
	If $X_1, X_2, \ldots, X_n$ are a *sequence* of i.i.d. random variables with finite $$E[X_1] = E[X_2] = \cdots = E[X_n] = \mu$$ Then for any constant $k > 0$, $$P\left(\left|\frac{X_1 + X_2 + \cdots + X_n}{n} - \mu\right| \geq k\right) \to 0$$ *as* $n \to \infty$

7.4 Limit Theorems in Probability

Table 7.8 provides the definitions of the basic limit theorems in probability.

Example 7 The length of the pencils in a batch is defined as a random variable X with $E[X] = \mu = 17$ and $Var(X) = \sigma^2 = 5$. The upper-bound probability that the length of a randomly selected pencil is greater than 18 cm will be

$$P(X \geq 18) \leq \frac{\mu}{18} = \frac{17}{18} = 0.9444$$

by *Markov's inequality*.

The upper-bound probability that the length of a randomly selected pencil is greater than 27 cm or less than 7 cm will be

$$P(X > 27) + P(X < 7)$$
$$= P(|X - 17| \geq 10) \leq \frac{\sigma^2}{(10)^2} = \frac{5}{100} = 0.05$$

by *Chebyshev's inequality.*

Example 8 A random experiment of rolling *three dice* is performed *independently* for 1,000 times. Let X_i : Number of outcomes with prime numbers for trial i, $1 \leq i \leq 1,000$.

Then the average number of outcomes with prime numbers will be

$$\frac{X_1 + X_2 + \cdots + X_{1,000}}{1,000} \to \mu$$

by *Strong Law of Large Numbers.* Note that for all $1 \leq i \leq 1,000$,

$$\mu = E[X_i] = (0)P(X_i = 0) + (1)P(X_i = 1) + (2)P(X_i = 2) + (3)P(X_i = 3)$$
$$= P(X_i = 1) + 2P(X_i = 2) + 3P(X_i = 3)$$

$$= \left(\frac{\binom{3}{1} 3.3.3}{6.6.6}\right) + 2\left(\frac{\binom{3}{2} 3.3.3}{6.6.6}\right) + 3\left(\frac{\binom{3}{3} 3.3.3}{6.6.6}\right)$$

$$= \left(\frac{3}{8}\right) + 2\left(\frac{3}{8}\right) + 3\left(\frac{1}{8}\right) = \frac{12}{8} = \frac{3}{2} = 1.5$$

Example 8 (revisited 1) The probability that the total number of outcomes with prime numbers is more than 1,550 will be

$$P\left(X_1 + X_2 + \cdots + X_{1,000} > 1,550\right)$$

The *approximated* probability can be computed by using *Central Limit Theorem* such that

$$P\left(\frac{X_1 + X_2 + \cdots + X_{1,000} - 1,000E[X_i]}{\sqrt{1,000Var(X_i)}} > \frac{1,550 - 1,000E[X_i]}{\sqrt{1,000Var(X_i)}}\right)$$
$$= P\left(Z > \frac{1,550 - 1,000E[X_i]}{\sqrt{1,000Var(X_i)}}\right)$$

Note that

$$E[X_i] = \frac{3}{2}$$

from Example 8, and

$$Var(X_i) = \sigma^2 = E[X_i^2] - (E[X_i])^2$$

where

$$E[X_i^2] = (1^2)\left(\frac{3}{8}\right) + (2^2)\left(\frac{3}{8}\right) + (3^2)\left(\frac{1}{8}\right) = 3$$

and

$$Var(X_i) = \sigma^2 = 3 - \left(\frac{3}{2}\right)^2 = \frac{3}{4}$$

Finally,

$$P\left(Z > \frac{1{,}550 - 1{,}000\left(\frac{3}{2}\right)}{\sqrt{1{,}000\left(\frac{3}{4}\right)}}\right) = P\left(Z > \frac{50}{\sqrt{750}}\right)$$

$$= P(Z > 1.8257) = 1 - \Phi(1.8257) \cong 1 - 0.9664 = 0.0336$$

Problems

1. Let X be a *geometric* random variable with parameter p. Find $E[X]$ and $Var(X)$ by using the moment generating function.

Solution
Since

$$M_X(t) = \frac{pe^t}{1 - (1-p)e^t} \quad (1-p)e^t < 1$$

is the moment generating function of a geometric random variable X with parameter p from Table 7.7,

$$M_X'(t) = \frac{pe^t\left(1 - (1-p)e^t\right) + pe^t(1-p)e^t}{(1-(1-p)e^t)^2} = \frac{pe^t}{(1-(1-p)e^t)^2}$$

$$M_X'(0) = \frac{p}{p^2} = \frac{1}{p} = E[X]$$

$$M_X''(t) = \frac{pe^t\left(1 - (1-p)e^t\right)^2 + 2p(1-p)e^{2t}\left(1 - (1-p)e^t\right)}{(1-(1-p)e^t)^4}$$

$$M_X''(0) = \frac{pp^2 + 2p^2(1-p)}{p^4} = \frac{1}{p} + \frac{2}{p^2} - \frac{2}{p} = \frac{2}{p^2} - \frac{1}{p} = E[X^2]$$

$$Var(X) = E[X^2] - (E[X])^2 = \left(\frac{2}{p^2} - \frac{1}{p}\right) - \frac{1}{p^2} = \frac{1}{p^2} - \frac{1}{p} = \frac{1-p}{p^2}$$

2. Two *independent* random experiments are to be performed simultaneously. A coin with heads probability $\frac{1}{3}$ is tossed 15 times *independently* from each other. An urn contains 4 red balls, 5 green balls, and 3 black balls, and a ball is selected from this urn 10 times *with replacement* and *independently* from each other. Find the probability that the total number of outcomes with heads and red balls will be 8 as a result of performing these two random experiments.

Solution
Let

X: Number of outcomes with *heads* for the first experiment
Y: Number of outcomes with *red* balls for the second experiment

By noting that the probability of selecting a red ball is $\frac{4}{4+5+3} = \frac{1}{3}$ for each withdrawal,

$$X \sim Bin\left(15, \frac{1}{3}\right)$$

$$Y \sim Bin\left(10, \frac{1}{3}\right)$$

Accordingly,

$$X + Y \sim Bin\left(15 + 10, \frac{1}{3}\right) \text{ and } X + Y \sim Bin\left(25, \frac{1}{3}\right)$$

by the *sum of independent binomial* random variables from Table 7.3, and the probability that the total number of the outcomes with heads and red balls will be 8 will be

$$P(X + Y = 8) = \binom{25}{8}\left(\frac{1}{3}\right)^8\left(1 - \frac{1}{3}\right)^{25-8} = \binom{25}{8}\left(\frac{1}{3}\right)^8\left(\frac{2}{3}\right)^{17} = 0.1673$$

The binomial random variables X and Y can also be *approximated* to the Poisson random variables such that

$$X \sim Poiss(5)$$

$$Y \sim Poiss\left(\frac{10}{3}\right)$$

Accordingly,

$$(X + Y) \sim Poiss\left(5 + \frac{10}{3}\right) \text{ and} (X + Y) \sim Poiss\left(\frac{25}{3}\right)$$

by the *sum of independent Poisson* random variables from Table 7.3, and the probability that the total number of the outcomes with heads and red balls will be 8 will be

$$P(X + Y = 8) = \frac{e^{-\frac{25}{3}} \left(\frac{25}{3}\right)^8}{8!} = 0.1386$$

3. Recall the joint probability density function of the continuous random variables X and Y as

$$f(x, y) = x + y, \quad 0 < x < 1, \quad 0 < y < 1$$

from Example 2 in this chapter. Find the expected value of X by using the identity $E[X] = E[E(X|Y)]$.

Solution
Recall from Example 2

$$f_Y(y) = y + \frac{1}{2}$$

and from Example 2 (revisited 1)

$$E(X|Y = y) = \frac{1}{2y + 1}\left(\frac{2}{3} + y\right)$$

Hence,

$$E[X] = E[E(X|Y)] = \int_0^1 E(X|Y = y) f_Y(y) dy$$

$$= \int_0^1 \frac{1}{2y + 1}\left(\frac{2}{3} + y\right)\left(y + \frac{1}{2}\right) dy$$

$$= \int_0^1 \left(\frac{1}{3} + \frac{y}{2}\right) dy = \frac{7}{12} = E[X]$$

Please recall from Example 2 also that

$$f_X(x) = x + \frac{1}{2}$$

We could find the expected value of X also by

$$E[X] = \int_0^1 x f_X(x)dx = \int_0^1 x\left(x + \frac{1}{2}\right)dx = \int_0^1 \left(x^2 + \frac{1}{2}x\right)dx = \frac{7}{12}$$

4. If it is rainy, the number of accidents on a given road is *Poisson* distributed with mean 4; and if it is not rainy, the number of accidents on that given road is *Poisson* distributed with mean 1. The probability that it will be rainy tomorrow is 0.6. Find the expected value and the variance of the number of accidents for tomorrow.

Solution

Let

X: Number of accidents on that given road for tomorrow
R: Event that it is rainy tomorrow

Note that

$$E(X|R) = Var(X|R) = 4$$
$$E(X|R^C) = Var(X|R^C) = 1$$
$$P(R) = 0.6$$

Accordingly, the expected value of the number of accidents for tomorrow will be

$$E[X] = E(X|R)P(R) + E(X|R^C)P(R^C)$$
$$= 4(0.6) + 1(1 - 0.6) = 2.8$$

and the variance of the number of accidents for tomorrow will be

$$Var(X) = \sigma^2 = E[X^2] - (E[X])^2$$

where

$$E[X^2] = E(X^2|R)P(R) + E(X^2|R^C)P(R^C)$$
$$= 20(0.6) + 2(1 - 0.6) = 12.8$$

since

$$E(X^2|R) = Var(X|R) + (E(X|R))^2$$
$$= 4 + (4)^2 = 20$$
$$E(X^2|R^C) = Var(X|R^C) + (E(X|R^C))^2$$
$$= 1 + (1)^2 = 2$$

Finally,

$$Var(X) = \sigma^2 = E[X^2] - (E[X])^2 = 12.8 - (2.8)^2 = 4.96$$

5. Let X_1 and X_2 be *independent exponential* random variables with rates λ_1 and λ_2. Find the probability density function of $Y = \frac{X_1}{X_2}$ for $X_2 > 0$.

Solution
We start with determining the cumulative distribution function of Y by conditioning on X_2 such that

$$
\begin{aligned}
F_Y(y) &= \int_0^\infty P(Y \le y | X_2 = x) f_{X_2}(x) dx \\
&= \int_0^\infty P\left(\frac{X_1}{X_2} \le y | X_2 = x\right) f_{X_2}(x) dx \\
&= \int_0^\infty P(X_1 \le xy | X_2 = x) f_{X_2}(x) dx \\
&= \int_0^\infty P(X_1 \le xy) f_{X_2}(x) dx \\
&= \int_0^\infty (1 - e^{-\lambda_1 xy}) \lambda_2 e^{-\lambda_2 x} dx \\
&= \int_0^\infty \lambda_2 e^{-\lambda_2 x} dx - \int_0^\infty \lambda_2 e^{-(\lambda_1 y + \lambda_2)x} dx \\
&= 1 - \frac{\lambda_2}{\lambda_1 y + \lambda_2} = \frac{\lambda_1 y}{\lambda_1 y + \lambda_2}
\end{aligned}
$$

Since

$$\frac{dF_Y(y)}{d(y)} = f_Y(y)$$

$$f_Y(y) = \left(\frac{\lambda_1 y}{\lambda_1 y + \lambda_2}\right)' = \frac{\lambda_1(\lambda_1 y + \lambda_2) - \lambda_1 y \lambda_1}{(\lambda_1 y + \lambda_2)^2} = \frac{\lambda_1 \lambda_2}{(\lambda_1 y + \lambda_2)^2}$$

Remark 6 Note that $\int_0^\infty P(X_1 \le xy | X_2 = x) f_{X_2}(x) dx$ $=$
$\int_0^\infty P(X_1 \le xy) f_{X_2}(x) dx$ holds, since X_1 and X_2 are independent.

6. An assembly worker has to assemble 50 parts. The time to assemble each part is *i.i.d.* with mean 8 min, and standard deviation 2 min. *Approximate* the probability that all 50 parts will be assembled in less than 7 h.

Solution

Let X_i : Time to assemble part i, $i = 1, 2, \ldots, 50$
Then the question will be

$$P(X_1 + X_2 + \cdots + X_{50} < 420)$$

By considering *Central Limit Theorem*,

$$P\left(\frac{X_1 + X_2 + \cdots + X_{50} - E[X_1 + X_2 + \cdots + X_{50}]}{\sqrt{Var(X_1 + X_2 + \cdots + X_{50})}} < \frac{420 - E[X_1 + X_2 + \cdots + X_{50}]}{\sqrt{Var(X_1 + X_2 + \cdots + X_{50})}} \right)$$

$$= P\left(Z < \frac{420 - E[X_1 + X_2 + \cdots + X_{50}]}{\sqrt{Var(X_1 + X_2 + \cdots + X_{50})}} \right)$$

Note that since X_i's are *i.i.d.*,

$$E[X_i] = 8$$
$$Var(X_i) = (2)^2 = 4$$

hold for all $i = 1, 2, \ldots, 50$, and

$$E[X_1 + X_2 + \cdots + X_{50}] = 50E[X_i] = 50(8) = 400$$
$$Var(X_1 + X_2 + \cdots + X_{50}) = 50Var(X_i) = 50(4) = 200$$

Finally,

$$P\left(Z < \frac{420 - 400}{\sqrt{200}} \right) = P\left(Z < \frac{20}{10\sqrt{2}} \right)$$
$$= P\left(Z < \sqrt{2} \right) = \Phi\left(\sqrt{2} \right) = \Phi(1.4142) \cong 0.9207$$

7. The number of the typographical errors on a book page is assumed to be *independent* and *Poisson* distributed with mean 2 for each page. If we select 40 pages from this book randomly, find the probability that there are at most 100 errors. Find this probability also by *approximation*.

Solution

Let X_i: Number of the typographical errors on a randomly selected page i, $i = 1, 2, \ldots, 40$

Since

$$X_i \sim Poiss(2)$$

holds for all $i = 1, 2, \ldots, 40$ *independently* from each other, it follows that

$$\left(\sum_{i=1}^{40} X_i\right) \sim Poiss((2)(40)) \text{ and } \left(\sum_{i=1}^{40} X_i\right) \sim Poiss(80)$$

and the probability that there are at most 100 errors on the randomly selected 40 pages will be

$$P\left(\sum_{i=1}^{40} X_i \leq 100\right) = \sum_{k=0}^{100} \frac{e^{-80}(80)^k}{k!} = 0.9868$$

Since $X_i's$ are *i.i.d.*, the *approximated* probability that there are at most 100 errors will also be

$$P\left(\frac{X_1 + X_2 + \cdots + X_{40} - E[X_1 + X_2 + \cdots + X_{40}]}{\sqrt{Var(X_1 + X_2 + \cdots + X_{40})}} \leq \frac{100 - E[X_1 + X_2 + \cdots + X_{40}]}{\sqrt{Var(X_1 + X_2 + \cdots + X_{40})}}\right)$$

$$= P\left(Z \leq \frac{100 - E[X_1 + X_2 + \cdots + X_{40}]}{\sqrt{Var(X_1 + X_2 + \cdots + X_{40})}}\right)$$

by *Central Limit Theorem*, where

$$E[X_1 + X_2 + \cdots + X_{40}] = (2)(40) = 80$$
$$Var(X_1 + X_2 + \cdots + X_{40}) = (2)(40) = 80$$

Accordingly,

$$P\left(Z \leq \frac{100 - 80}{\sqrt{80}}\right) = P\left(Z \leq \frac{20}{4\sqrt{5}}\right) = P\left(Z \leq \sqrt{5}\right)$$

$$= \Phi\left(\sqrt{5}\right) = \Phi(2.2361) \cong 0.9875$$

8. A die is rolled *three* times *independently* from each other. Let X be the sum of the outcomes, and let Y be the first outcome plus second outcome minus third outcome. Compute

    (a) $Cov(X, Y)$
    (b) $Corr(X, Y)$

Solution

Let Z_i: Outcome of the ith roll, $i = 1, 2, 3$

Hence,

$$X = Z_1 + Z_2 + Z_3$$
$$Y = Z_1 + Z_2 - Z_3$$

(a)

$$
\begin{aligned}
Cov(X, Y) &= Cov(Z_1 + Z_2 + Z_3, Z_1 + Z_2 - Z_3) \\
&= Cov(Z_1, Z_1) + Cov(Z_1, Z_2) - Cov(Z_1, Z_3) \\
&\quad + Cov(Z_2, Z_1) + Cov(Z_2, Z_2) - Cov(Z_2, Z_3) \\
&\quad + Cov(Z_3, Z_1) + Cov(Z_3, Z_2) - Cov(Z_3, Z_3) \\
&= Var(Z_1) + 2Cov(Z_1, Z_2) + Var(Z_2) - Var(Z_3) \\
&= Var(Z_1) = \frac{105}{36} = 2.9167
\end{aligned}
$$

Note that $Cov(Z_1, Z_2) = 0$ since Z_1 and Z_2 are *independent* random variables, $Var(Z_2) = Var(Z_3)$ since the 2nd and the 3rd rolls are *independent*, and

$$
\begin{aligned}
Var(Z_1) &= E[Z_1^2] - (E[Z_1])^2 \\
&= \frac{91}{6} - \left(\frac{21}{6}\right)^2 = \frac{105}{36} = 2.9167
\end{aligned}
$$

where

$$
\begin{aligned}
E[Z_1^2] &= 1^2\left(\frac{1}{6}\right) + 2^2\left(\frac{1}{6}\right) + 3^2\left(\frac{1}{6}\right) \\
&\quad + 4^2\left(\frac{1}{6}\right) + 5^2\left(\frac{1}{6}\right) + 6^2\left(\frac{1}{6}\right) = \frac{91}{6} \\
E[Z_1] &= 1\left(\frac{1}{6}\right) + 2\left(\frac{1}{6}\right) + 3\left(\frac{1}{6}\right) + 4\left(\frac{1}{6}\right) + 5\left(\frac{1}{6}\right) \\
&\quad + 6\left(\frac{1}{6}\right) = \frac{21}{6}
\end{aligned}
$$

(b) $Corr(X, Y) = \dfrac{Cov(X, Y)}{\sqrt{Var(X)Var(Y)}}$

where

$$Cov(X, Y) = Var(Z_1) = \frac{105}{36}$$

from (a),

$$Var(X) = Var(Z_1 + Z_2 + Z_3) = Var(Z_1) + Var(Z_2) + Var(Z_3)$$
$$= 3Var(Z_1) = 3\left(\frac{105}{36}\right) = \frac{315}{36}$$
$$Var(Y) = Var(Z_1 + Z_2 - Z_3) = Var(Z_1) + Var(Z_2) + Var(Z_3)$$
$$= 3Var(Z_1) = \frac{315}{36}$$

Finally,

$$Corr(X, Y) = \frac{Cov(X, Y)}{\sqrt{Var(X)Var(Y)}} = \frac{\left(\frac{105}{36}\right)}{\sqrt{\left(\frac{315}{36}\right)\left(\frac{315}{36}\right)}} = 0.3333$$

9. A die is rolled twice *independently* from each other. Let X_1 and X_2 be the outcomes of the first roll and the second roll, respectively. Find the probability mass function of $X_1 + X_2$.

Solution
By considering the following formula for the *convolution* of p_{X_1} and p_{X_2} for the *discrete* and *independent* random variables X_1 and X_2,

$$p_{X_1+X_2}(a) = \sum_{x_2} p_{X_1}(a - x_2)p_{X_2}(x_2)$$

we can write the probability mass function of $X_1 + X_2$ as follows:

$$p_{X_1+X_2}(2) = p_{X_1}(1)p_{X_2}(1) = \left(\frac{1}{6}\right)\left(\frac{1}{6}\right) = \frac{1}{36}$$

$$p_{X_1+X_2}(3) = p_{X_1}(1)p_{X_2}(2) + p_{X_1}(2)p_{X_2}(1) = \left(\frac{1}{6}\right)\left(\frac{1}{6}\right) + \left(\frac{1}{6}\right)\left(\frac{1}{6}\right) = \frac{2}{36}$$

$$p_{X_1+X_2}(4) = p_{X_1}(1)p_{X_2}(3) + p_{X_1}(2)p_{X_2}(2) + p_{X_1}(3)p_{X_2}(1)$$
$$= \left(\frac{1}{6}\right)\left(\frac{1}{6}\right) + \left(\frac{1}{6}\right)\left(\frac{1}{6}\right) + \left(\frac{1}{6}\right)\left(\frac{1}{6}\right) = \frac{3}{36}$$

$$p_{X_1+X_2}(5) = p_{X_1}(1)p_{X_2}(4) + p_{X_1}(2)p_{X_2}(3) + p_{X_1}(3)p_{X_2}(2) + p_{X_1}(4)p_{X_2}(1)$$
$$= \left(\frac{1}{6}\right)\left(\frac{1}{6}\right) + \left(\frac{1}{6}\right)\left(\frac{1}{6}\right) + \left(\frac{1}{6}\right)\left(\frac{1}{6}\right) + \left(\frac{1}{6}\right)\left(\frac{1}{6}\right) = \frac{4}{36}$$

Analogously,

$$p_{X_1+X_2}(6) = \frac{5}{36}$$

$$p_{X_1+X_2}(7) = \frac{6}{36}$$

$$p_{X_1+X_2}(8) = \frac{5}{36}$$

$$p_{X_1+X_2}(9) = \frac{4}{36}$$

$$p_{X_1+X_2}(10) = \frac{3}{36}$$

$$p_{X_1+X_2}(11) = \frac{2}{36}$$

$$p_{X_1+X_2}(12) = \frac{1}{36}$$

10. Let X_1, X_2, X_3 be *independent* and *uniform* random variables defined over $(0, 1)$.

   (a) Find the joint probability density function of the order statistics of X_1, X_2, X_3.
   (b) Find $P\left(X_2 + X_3 \leq \frac{3}{2}\right)$.
   (c) Find $P(X_1 + X_2 + X_3 \leq 1)$.

Solution

(a) Recall from Example 4 in this chapter that for independent $X_i \sim Uniform(\alpha, \beta), i = 1, 2, \ldots, n$ random variables, the joint probability density function of $X_{(1)}, X_{(2)}, X_{(3)}, \cdots, X_{(n)}$ is

$$f_{X_{(1)}, X_{(2)}, \ldots, X_{(n)}}(x_1, x_2, \ldots, x_n) = n!\left(\frac{1}{\beta - \alpha}\right)^n \quad \text{for } \alpha < x_1 < x_2 < \ldots < x_n < \beta$$

Accordingly, for independent $X_i \sim Uniform(0, 1), i = 1, 2, 3$ random variables, the joint probability density function of $X_{(1)}, X_{(2)}, X_{(3)}$ will be

$$f_{X_{(1)}, X_{(2)}, X_{(3)}}(x_1, x_2, x_3) = 3!(1)^3 = 6 \quad \text{for } 0 < x_1 < x_2 < x_3 < 1$$

(b) Since $X_2 + X_3$ is a *triangular random variable*,

$$P\left(X_2 + X_3 \leq \frac{3}{2}\right) = \int\limits_{0}^{1} x\,dx + \int\limits_{1}^{3/2} (2 - x)\,dx = \frac{7}{8}$$

(c) pagebreakSince $X_1 + X_2 + X_3$ is *Irwin-Hall* distributed, it follows that

$$P(X_1 + X_2 + X_3 \leq 1) = F(1) = \sum_{i=0}^{3}\left[(-1)^i\binom{3}{i}\frac{(1-i)^3}{3!}s_i(1)\right]$$

$$= \left[(-1)^0\binom{3}{0}\frac{(1-0)^3}{3!}s_0(1)\right] + \left[(-1)^1\binom{3}{1}\frac{(1-1)^3}{3!}s_1(1)\right]$$

$$+ \left[(-1)^2\binom{3}{2}\frac{(1-2)^3}{3!}s_2(1)\right] + \left[(-1)^3\binom{3}{3}\frac{(1-3)^3}{3!}s_3(1)\right]$$

$$= \frac{1}{3!} = \frac{1}{6}$$

where

$$s_i(1) = \begin{cases} 0 \, if \, i > 1 \\ 1 \, if \, i \leq 1 \end{cases}$$

is the unit step function and

$$s_0(1) = s_1(1) = 1$$
$$s_2(1) = s_3(1) = 0$$

11. Let X_1, X_2, X_3 be *independent* continuous random variables with common probability density function $f(x) = 2x$ for $0 < x < 1$. Find the joint probability density function of the order statistics of X_1, X_2, X_3.

Solution
The joint probability density function of the order statistics of X_1, X_2, X_3 will be

$$f_{X_{(1)}, X_{(2)}, X_{(3)}}(x_1, x_2, x_3) = 3!(2x_1)(2x_2)(2x_3) = 48x_1x_2x_3$$

where $0 < x_1 < x_2 < x_3 < 1$.

Part II
Basics of Stochastic Processes

Chapter 8
A Brief Introduction to Stochastic Processes

Abstract This chapter was a very brief introduction to the basic concepts of the stochastic processes including the definition of a stochastic process, a discrete-time and continuous-time stochastic process, state space, *i.i.d.* stochastic process, stopping time, and the hitting time of a process to a state. Some typical examples and problems including the daily stock prices of a company, and the evolution of the population of a country have been provided to clarify the basic concepts.

Although random variables provided in Chaps. 4, 5, and 6, and jointly distributed random variables provided in Chap. 7 are very important in their own sense; they do not consider the evolution of a random phenomenon with respect to time. However, many random phenomena should be modelled with respect to time to be able to foresee the evolution in the future. This chapter will be a very brief introduction to the stochastic processes with the basic concepts including the definition of a stochastic process, a discrete-time and continuous-time stochastic process, state space, *i.i.d.* stochastic process, stopping time, and hitting time of a process to a state. As it is clear from Chap. 7, the assumption of *i.i.d.* random variables is already important in probability, for example in the order statistics of the random variables, and in the limit theorems including Central Limit Theorem and Strong Law of Large Numbers. Its importance for the stochastic processes will also be clear in the remaining chapters.

Example 1 Think about the daily weather conditions in a region for the following one year. It is clear that since we consider the future, the weather condition for each day can be defined as a *random variable*. Thus, we have *a series (collection) of random variables* that represent the daily weather conditions for the following year.

Definition 1 (*Stochastic process*) A stochastic process is *a series (collection) of random variables* that describes the *evolution* of a random phenomenon generally with respect to *time*. A stochastic process can be *discrete-time* denoted as

$$\{X_n; \quad n \geq 0\}$$

or *continuous-time* denoted as

$$\{X(t); \quad t \geq 0\}$$

© Springer Nature Switzerland AG 2019

E. Bas, *Basics of Probability and Stochastic Processes*,
https://doi.org/10.1007/978-3-030-32323-3_8

Example 1 (revisited 1) The stochastic process in Example 1 can be defined as

$$\{X_n; \quad n = 0, 1, 2, \dots, 365\}$$

where $n = 0$ represents the *initial day (today)*, and X_n: Random variable that represents the weather condition on the nth day.

Since this stochastic process represents the *daily* weather conditions; time is *discrete*, and it is a *discrete-time* stochastic process.

Example 2 The *evolution of temperature* of a chemical process can be represented by using a *continuous-time* stochastic process $\{X(t); t \geq 0\}$.

Definition 2 (*State space*) The set of all possible states of a random variable of a stochastic process is called as a *state space*, and denoted by S. A stochastic process can be a *discrete-state space* or *continuous-state space* process.

Example 1 (revisited 2) The state space for the random variables of this stochastic process can be defined as $S = \{1, 2, 3, 4, 5\}$, where

1: Sunny
2: Rainy
3: Snowy
4: Windy
5: Cloudy.

Definition 3 (*i.i.d. stochastic process*)
    A stochastic process is *independent and identically distributed* (*i.i.d.*), if its random variables are *i.i.d.*

Example 1 (revisited 3) If the random variables representing the daily weather conditions are assumed to be *independent*, and with the same probability values $p(i), \ i = 1, 2, 3, 4, 5$ for each day, then this stochastic process will be an *i.i.d.* stochastic process.

Definition 4 (*Stopping time*) Let $\{X_n; n \geq 0\}$ be a *discrete-time* stochastic process. A *stopping time* τ with respect to this stochastic process can be defined as a *random time* n^*, which can be determined at most by the states $\{X_0, X_1, X_2, \dots, X_{n^*}\}$. Let $\{X(t); t \geq 0\}$ be a *continuous-time* stochastic process. A *stopping time* τ with respect to this stochastic process can be defined as a *random time* t^*, which can be determined at most by the states $\{X(t); 0 \leq t \leq t^*\}$.

Definition 5 (*Hitting time of a process to a state, first passage time of a process into a state*) Table 8.1 provides the definition of the hitting time of a discrete-time and continuous-time stochastic process to a state.

Remark 1 It can be shown that a *hitting time* of a process to a state is also a *stopping time*.

Table 8.1 Hitting time of a process to a state

Discrete-time stochastic process	Continuous-time stochastic process
$N_i = \min\{n \geq 0 \mid X_n = i\}$ can be defined as a *hitting time* of the discrete-time stochastic process to *state i* or the first passage time of the process into *state i*.	$T_i = \min\{t \geq 0 \mid X(t) = i\}$ can be defined as a *hitting time* of the continuous-time stochastic process to *state i* or the first passage time of the process into *state i*.
More generally, $N_A = \min\{n \geq 0 \mid X_n \in A\}$ can be defined as a *hitting time* of the discrete-time stochastic process to *set A* or the first passage time of the process into *set A*.	More generally, $T_A = \min\{t \geq 0 \mid X(t) \in A\}$ can be defined as a *hitting time* of the continuous-time stochastic process to *set A* or the first passage time of the process into *set A*.

Example 1 (revisited 4) A *stopping time* for the stochastic process of the weather conditions can be defined as τ, where τ is the first *snowy* day starting today. As an example, the weather conditions $X_0, X_1, \ldots, X_7$ determine whether $\tau = 7$ holds. By recalling from Example 1 (revisited 2) that *state* 3 denotes the snowy day, the first *snowy* day can also be defined as a *hitting time of the process to state* 3 such that

$$N_3 = \min\{n \geq 0 \mid X_n = 3\}$$

As another example, the first *rainy* or *snowy* day can also be defined as a *hitting time of the process to set A* such that

$$N_A = \min\{n \geq 0 \mid X_n \in A\}$$

where $A \in \{2, 3\}$, by recalling that *state* 2 denotes the rainy day, and *state* 3 denotes the snowy day.

Problems

1. Define the following stochastic processes as *discrete-time* or *continuous-time* stochastic processes.

   (a) Daily stock prices of a company
   (b) The evolution of the blood pressure values of a patient in a surgery
   (c) The evolution of the population of a country.

Answer

(a) If the stock price values are assumed to change *daily*, then this stochastic process is a *discrete-time stochastic process*.
(b) If the blood pressure values of a patient in a surgery are assumed to change *continuously*, then this stochastic process is a *continuous-time stochastic process*.
(c) If it can be assumed that there will be births and deaths at each second in a country, then this stochastic process is also a *continuous-time stochastic process*.

2. Define the following phenomena as stochastic processes, define their state spaces, and define a hitting time for each stochastic process.

(a) An infinitely performed random experiment of withdrawing two balls *with replacement* from an urn including 5 white balls, 7green balls, and 2 red balls.
(b) Number of the typographical errors on the pages of a book with 350 pages
(c) The evolution of the systolic blood pressure values of a patient in a surgery

Answer

(a) A *discrete-time* stochastic process can be defined as

$$\{X_n; \quad n \geq 1\}$$

where

X_n : Outcome of the nth withdrawal

with *discrete-state space*

$$S = \{1, 2, 3, 4, 5, 6\}$$

$$1 : WW$$
$$2 : GG$$
$$3 : RR$$
$$4 : WG$$
$$5 : WR$$
$$6 : GR$$

The number of the withdrawals until the outcome with one *white* ball and one *green* ball can be defined as a hitting time of the process to *state* 4 such that

$$N_4 = min\{n \geq 1 | X_n = 4\}$$

The number of the withdrawals until the outcome including *at least one red* ball can also be defined as a hitting time of the process to *set R* such that

$$N_R = min\{n \geq 1 | X_n \in R\}$$

where $R = \{3, 5, 6\}$.

(b) A *discrete-time* stochastic process can be defined as

$$\{X_n; \quad n = 1, 2, \ldots, 350\}$$

where

X_n : Number of the typographical errors on the nth page

with *discrete-state space* for the nth page as

$$S_n = \big[0, \text{number of the letters on the } n\text{th page}\big]$$

The *page number* of the first page with at least 5 errors can be defined as a hitting time of the process to *set A* such that

$$N_A = min\{n \geq 1 | X_n \in A\}$$

where $A = \{5, 6, 7, \ldots\}$.

(c) A *continuous-time* stochastic process can be defined as

$$\{X(t); \quad t \geq 0\}$$

where

$X(t)$: Systolic blood pressure value of the patient at time t

with *continuous-state space*

$$S = [70, 190]$$

The blood pressure value less than 80 can be defined as a hitting time of the process to set B such that

$$T_B = min\{t \geq 0 | X(t) \in B\}$$

where $B \in [70, 80)$.

Chapter 9
A Brief Introduction to Point Process, Counting Process, Renewal Process, Regenerative Process, Poisson Process

Abstract This chapter started with the definition of point process as a basic stochastic process, continued with the definitions of the arrival time and the interarrival time; and based on the definition of point process, counting process has been defined. The definition of renewal process as a special counting process, and the definition of Poisson process as a special renewal process have been provided. Poisson process has been classified as homogeneous and nonhomogeneous Poisson process, and the basic properties of a homogeneous Poisson process including the stationary and independent increments, exponential random variable as the interarrival time distribution, and gamma random variable as the arrival time distribution have been explained. In addition to the typical example of the arrival of the customers; other illustrative examples including the breakdown of the machines, time of the earthquakes, and the two-server systems have also been presented, and solved by using the basic formulae and the schematic representations.

This chapter is a brief introduction to the basic stochastic processes in a systematic manner. Although modelling the evolution of many random phenomena such as the evolution of the earthquake times with Poisson process is very common; in this chapter, it is highlighted that homogeneous *Poisson process* is a special *renewal process* with *exponentially* distributed interarrival times, *renewal process* is a special *counting process* with *i.i.d.* interarrival times, and *counting process* can be defined based on *point process*. *Poisson* random variable provided in Chap. 5, and *exponential* random variable and *gamma* random variable provided in Chap. 6 are important in this chapter and in Chap. 10. The examples and problems presented in this chapter will be a solid background for Chap. 10. Since some problems are related to the renewal process and the server systems, these will also build some background for the renewal processes in Chap. 11 and for the queueing models in Chap. 15.

© Springer Nature Switzerland AG 2019
E. Bas, *Basics of Probability and Stochastic Processes*,
https://doi.org/10.1007/978-3-030-32323-3_9

9.1 Point Process, Counting Process

Example 1 Suppose that a shop opens at 9:00 a.m. every day, and the arrival times of the customers to this shop are the basic consideration.

Definition 1 (*Point process*) A point process is a stochastic process with *a series of* random variables that represent the *arrival times* (*waiting times*) of an event, and can be denoted as

$$\{t_n; n \geq 0\}$$

where
t_n: Arrival time (waiting time) of the nth event, nth arrival time

$$t_0 < t_1 < t_2 < \cdots < t_n < \cdots$$

Point process is illustrated in Fig. 9.1.

Remark 1 Although the time points $t_1, t_2, t_3, \ldots, t_n$ are shown on the time line, actually their exact positions are unknown, since they are random variables. Thus, we actually mean *random point process* with each *point process*. In a point process, $t_0 = 0$ means that the current time is assumed to be 0, and no event is assumed to occur at $t_0 = 0$. Only *one* event is assumed to occur at each time point $t_1, t_2, t_3, \ldots, t_n, \ldots,$ thus, a point process is also called as a *simple point process*.

Definition 2 (*Interarrival time*)
Consider the point process $\{t_n; n \geq 0\}$, where t_n is the arrival time of the nth event.
Then

$$X_1 = t_1 - t_0$$
$$X_2 = t_2 - t_1$$
$$X_3 = t_3 - t_2$$
$$\cdots$$
$$X_n = t_n - t_{n-1}$$

are the 1st, 2nd, 3rd, ..., nth interarrival times, respectively, for the point process

$$\{t_n; n \geq 0\}$$

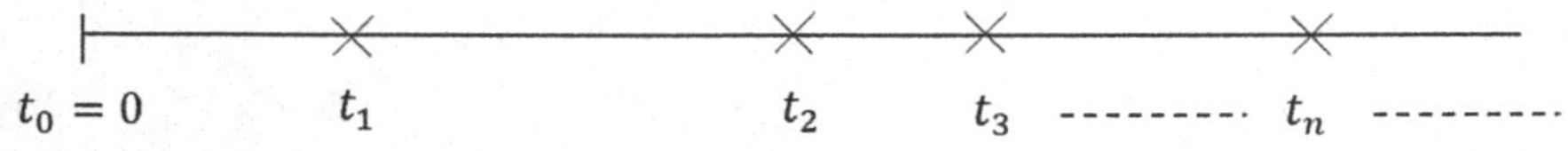

Fig. 9.1 Point process

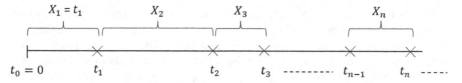

Fig. 9.2 Stochastic process of the interarrival times

and

$$\{X_n; n \geq 1\}$$

is a *stochastic process* with the random variables denoting the interarrival times. The stochastic process of the interarrival times is denoted in Fig. 9.2.

Note that

X_n: nth interarrival time and

$$t_n = X_1 + X_2 + \cdots + X_n$$

Definition 3 (*Counting process*)

A counting process is a stochastic process with *a series of random variables* that represent *the number of events* in a given time interval, and can be denoted as

$$\{N(t); t \geq 0\}$$

where $N(t)$: Number of events in the time interval $(0, t]$.

A counting process is illustrated in Fig. 9.3.

It is assumed that $N(0) = 0$, i.e. at $t = 0$, no event occurs. Note that

$$N(t) - N(s) : \text{Number of events in the time interval } (s, t] \text{ for } s < t$$

Example 1 (revisited 1) The arrival times of customers can be represented as a *point process* $\{t_n; n \geq 0\}$. As an example, t_{17} will be the arrival time of the 17th

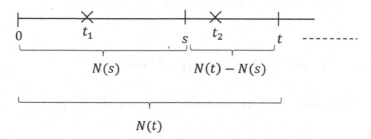

Fig. 9.3 Counting process

customer when we assume opening time 9:00 a.m. as $t = 0$. The *interarrival times* of customers can be represented as $\{X_n; n \geq 1\}$. As an example, X_8 will be the time between the arrivals of the 7th and the 8th customers. The stochastic process with the random variables denoting the number of customer arrivals in a time interval will be a *counting process* $\{N(t); t \geq 0\}$. As an example, $N(5) - N(3)$ will be the number of customer arrivals in the time interval $(3, 5]$, i.e. between 12:00 and 2:00 p.m. when we assume opening time 9:00 a.m. as $t = 0$.

Definition 4 (*Stationary increments and independent increments of a counting process*) *Stationary* increments occur if the distribution of the number of events depends only on the *length* of the time interval. As an example, in case of *stationary* increments, $N(t + s) - N(s)$ will have the same distribution with $N(t)$ for all $s \geq 0$, $t > 0$, since the *length* of the time interval is t for both increments.

   Independent increments occur if the number of events in *disjoint time intervals* is *independently* distributed. As an example, in case of *independent* increments, $N(t) - N(s)$ and $N(s)$ will be independently distributed for all $s \geq 0, t > 0$, since $(0, s]$ and $(s, t]$ are disjoint intervals as denoted in Fig. 9.3.

WARNING Not every counting process has *stationary* or *independent* increments. A counting process may have *both* stationary and independent increments, *or only* stationary increments, *or only* independent increments, *or* may have neither of them.

Example 1 (revisited 2) Recall that the shop opens at 9:00 a.m. If this counting process has *stationary increments*, then as an example

$$P(N(4) - N(2) \leq 3) = P(N(8) - N(6) \leq 3)$$
$$= P(N(2) - N(0) \leq 3) = P(N(2) \leq 3)$$

will hold since the time interval is 2 h for all increments.

   If this counting process has *independent increments*, then as an example

$$P(N(8) - N(5) = 4 | N(2) - N(1) = 5) = P(N(8) - N(5) = 4)$$

will hold since the time intervals $(1, 2]$ and $(5, 8]$ are *disjoint*.

9.2   Renewal Process, Regenerative Process

Definition 5 (*Renewal process*) A renewal process is a *counting process*

$$\{N(t); t \geq 0\}$$

in which the stochastic process of the interarrival times, i.e.,

$$\{X_n; n \geq 1\}$$

is an *i.i.d.* In a renewal process, we may replace the word *event* with the word *renewal*.

Example 1 (revisited 3) If the interarrival times of customers are *independent normal* random variables with the *same parameters* μ and σ^2, then this counting process will be a *renewal process*.

Definition 6 (*Regenerative process—Classical definition*)[1] A regenerative process $\{X(t); t \geq 0\}$ with state *i.i.d.* space S is a continuous-time stochastic process with the following properties:

There exists a random variable R_1 which implies the first *regeneration epoch* such that

(a) $\{X(t + R_1); t \geq 0\}$ is *independent* of $\{X(t); t \leq R_1\}$ and R_1
(b) $\{X(t + R_1); t \geq 0\}$ is *stochastically equivalent* to $\{X(t); t \leq R_1\}$.

Remark 2 According to the classical definition, the process beyond the first *regeneration epoch* R_1 is a *probabilistic replica* of the whole process. There are some other regeneration epochs $R_2, R_3, R_4, \ldots$ having the same property of R_1. All regeneration epochs $\{R_n; n \geq 1\}$ build also a stochastic process. Note that a *renewal process* is also a *regenerative* process with $\{t_n; n \geq 1\}$ as the sequence of the *regeneration epochs*.

9.3 Poisson Process

Definition 7 (*Poisson Process—PP*) A Poisson process is a *renewal process* so-called homogeneous PP in which the *interarrival* times $\{X_n; n \geq 1\}$ are *independent exponential* random variables with the same rate, i.e. if the parameter of a Poisson process is a constant λ for all time intervals, then this Poisson process is said to be a *homogeneous* Poisson process. If the parameter of a Poisson process is *time-dependent*, defined as *intensity function* $\lambda(t)$, then this Poisson process is said to be a *nonhomogeneous* Poisson process, which is a counting process but not a renewal process.

Definition 8 (*Homogeneous vs. nonhomogeneous Poisson process*) The basic properties of a homogeneous and nonhomogeneous Poisson process are provided in Table 9.1.

Example 1 (revisited 4) When the arrivals of customers are assumed to be according to a *homogeneous* Poisson process with parameter $\lambda = 10$ per hour, by assuming

---

[1] A review of regenerative processes by Sigman and Wolff.

Table 9.1 Homogeneous versus nonhomogeneous Poisson process

Homogeneous Poisson process	Nonhomogeneous Poisson process
• Independent increments	• Independent increments
• *Stationary* increments	• *Nonstationary* increments
• The number of events in a given time interval with any length is *Poisson* distributed with *parameter*	• The number of events in a given time interval is *Poisson* distributed with *intensity function*
$\lambda,\ \lambda > 0$	$\lambda(t)$
• The probability that k events will occur in the time interval $(s,\ s+t]$ will be	• The probability that k events will occur in the time interval $(s,\ s+t]$ will be
$$P(N(s+t) - N(s) = k) = P(N(t) = k)$$ $$= \frac{e^{-\lambda t}(\lambda t)^k}{k!}$$ and $$E[N(s+t) - N(s)] = E[N(t)] = \lambda t$$ $$Var(N(s+t) - N(s)) = Var(N(t)) = \lambda t$$	$$P(N(s+t) - N(s) = k)$$ $$= \frac{e^{-\int_s^{s+t} \lambda(y)dy}\left(\int_s^{s+t} \lambda(y)dy\right)^k}{k!}$$ and $$E[N(s+t) - N(s)] = \int_s^{s+t} \lambda(y)dy$$ $$Var(N(s+t) - N(s)) = \int_s^{s+t} \lambda(y)dy$$

9:00 a.m. as $t = 0$, the probability that 12 customers will arrive between 11:00 a.m. and 1:00 p.m. will be

$$P(N(4) - N(2) = 12) = P(N(2) = 12)$$
$$= \frac{e^{-(10)(2)}((10)(2))^{12}}{12!} = \frac{e^{-20}(20)^{12}}{12!} = 0.0176$$

due to the *stationary* increments.

Example 1 (revisited 5) If the arrivals of customers are assumed to be according to a *nonhomogeneous* Poisson process with *intensity function*

$$\lambda(t) = \begin{cases} 3t + 1 & 0 \leq t \leq 3 \\ 10 & 3 \leq t \leq 5 \\ 3t - 5 & 5 \leq t \leq 8 \end{cases}$$

then the probability that 12 customers will arrive between 11:00 a.m. and 1:00 p.m. will be

$$P(N(4) - N(2) = 12) = \frac{e^{-\left(\int_2^3 (3t+1)dt + \int_3^4 10dt\right)}\left(\int_2^3 (3t+1)dt + \int_3^4 10dt\right)^{12}}{12!}$$

$$= \frac{e^{-\left(\frac{37}{2}\right)}\left(\frac{37}{2}\right)^{12}}{12!} = 0.031$$

Fig. 9.4 Intensity function
$\lambda(t)$ for example 1 revisited 5

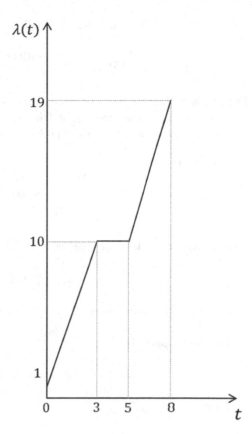

Please note that $E[N(4) - N(2)] = \frac{37}{2}$ holds, which can also be computed by finding the area under the function $\lambda(t)$ for $2 \le t \le 4$. The graph of $\lambda(t)$ is given in Fig. 9.4.

Example 1 (revisited 6) If the arrivals of customers are assumed to be according to a *homogeneous* Poisson process with parameter $\lambda = 10$ per hour, then the probability that "the number of customers who will arrive between 2:00 p.m. and 5:00 p.m. is 15" given that "the number of customers who arrive between 9:00 a.m. and 11:00 a.m. is 3" will be

$$P(N(8) - N(5) = 15 | N(2) = 3)$$
$$= P(N(8) - N(5) = 15)$$

by *independent* increments, since the time intervals $(0, 2]$ and $(5, 8]$ are *disjoint*, and

$$P(N(8) - N(5) = 15) = P(N(3) = 15)$$

will hold by *stationary* increments. Finally,

$$P(N(3) = 15) = \frac{e^{-(10)(3)}((10)(3))^{15}}{15!} = \frac{e^{-30}(30)^{15}}{15!} = 0.001$$

Definition 9 (*Interarrival time vs. arrival time distribution for a homogeneous Poisson process*) The interarrival time and the arrival time distribution for a homogeneous Poisson process with rate λ, and their basic formulae are provided in Table 9.2.

Remark 3 Let

$$X_1 \sim \exp(\lambda_1)$$
$$X_2 \sim \exp(\lambda_2)$$

be *independent* random variables, and let

$$Z = \min(X_1, X_2)$$

Accordingly, the following *further properties* of an exponential random variable will hold:

Further property 1

$$Z \sim \exp(\lambda_1 + \lambda_2)$$

Further property 2

$$P(X_1 < X_2) = \frac{\lambda_1}{\lambda_1 + \lambda_2}$$

Table 9.2 Interarrival time versus arrival time distribution for a homogeneous Poisson process

Interarrival time distribution for a homogeneous Poisson process	Arrival time (waiting time) distribution for a homogeneous Poisson process
Let $\{X_n; n \geq 1\}$ be the stochastic process of the *interarrival times* for a homogeneous Poisson process with rate λ. Hence, $X_n \sim \exp(\lambda), n \geq 1$	Let $\{t_n; n \geq 0\}$ be the stochastic process of the *arrival times* for a homogeneous Poisson process with rate λ. Hence, $t_n \sim Gamma(n, \lambda), n \geq 1$
Summary of the basic formulae of an exponential random variable	*Summary of the basic formulae of a gamma random variable*
$f(x) = \lambda e^{-\lambda x}, x \geq 0$ $F(x) = 1 - e^{-\lambda x}, x \geq 0$ $\overline{F}(x) = e^{-\lambda x}, x \geq 0$ $E[X] = \frac{1}{\lambda} \quad Var(X) = \frac{1}{\lambda^2}$	$f_{t_n}(t) = \frac{\lambda e^{-\lambda t}(\lambda t)^{n-1}}{(n-1)!}, t \geq 0$ $E[t_n] = \frac{n}{\lambda}, Var(t_n) = \frac{n}{\lambda^2}$

$$P(X_2 < X_1) = \frac{\lambda_2}{\lambda_1 + \lambda_2}$$

Example 1 (revisited 7) When the arrivals of customers are assumed to be according to a *homogeneous* Poisson process with rate $\lambda = 10$ per hour, the probability that the elapsed time between the arrivals of the 8th and the 9th customers exceeds 12 min will be

$$P\left(X_9 > \frac{12}{60}\right) = P\left(X_9 > \frac{1}{5}\right) = \bar{F}\left(\frac{1}{5}\right) = e^{-10(\frac{1}{5})} = 0.1353$$

where X_9: Elapsed time between the arrivals of the 8th and the 9th customers.
Since,

$$E[t_n] = \frac{n}{\lambda}$$

where $\lambda = 10$, the expected time until the 15th customer will be

$$E[t_{15}] = \frac{15}{10} = 1.5 \text{ h}$$

Example 2 There are two possible types of breakdown for a machine, called as type A and type B breakdown. The time until the first type A breakdown and the time until the first type B breakdown are *independent* and *exponentially* distributed with rates 3 and 4 per year, respectively. Hence, the time until the first breakdown will be

$$Z = \min(X_A, X_B)$$

where
  X_A: Time until the first type A breakdown
  X_B: Time until the first type B breakdown
  Accordingly,

$$Z \sim \exp(3+4) \text{ and } Z \sim \exp(7)$$

will hold and finally, the expected time until the first breakdown will be

$$E[Z] = \frac{1}{7} \text{ year} = 1.71 \text{ months}$$

Problems

1. Earthquakes are assumed to occur in a region according to a homogeneous Poisson process at a rate of $\frac{1}{4}$ per year.

(a)   What is the expected time until the 5th earthquake? What is the variance of
        the time until the 5th earthquake?
(b)   Given that 2 earthquakes occur in the following 4 years, what is the expected
        time until the 5th earthquake?

Solution

(a) Let

t_n: Time until the nth earthquake
Since,

$$E[t_n] = \frac{n}{\lambda}$$

$$Var(t_n) = \frac{n}{\lambda^2}$$

where $\lambda = \frac{1}{4}$,

$$E[t_5] = \frac{5}{\left(\frac{1}{4}\right)} = 20 \text{ years}$$

$$Var(t_5) = \frac{5}{\left(\frac{1}{4}\right)^2} = 80$$

(b)   We consider Fig. 9.5 for the solution.

$$E(t_5|N(4) = 2) = 4 + E[t_3] = 4 + \frac{3}{\left(\frac{1}{4}\right)} = 16 \text{ years}$$

Note that by *memoryless property of exponential* random variable, we may consider time $t = 4$ as the beginning, after which 3 events occur.

2. Let

$$\{N(t); t \geq 0\}$$

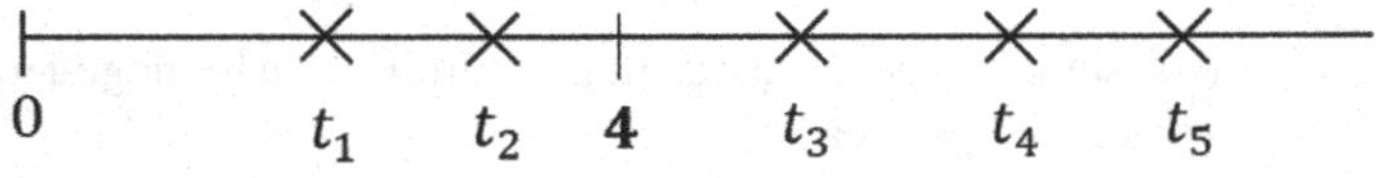

Fig. 9.5 Schematic representation of problem 1(b)

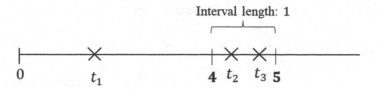

Fig. 9.6 Schematic representation of problem 2(b)

be a homogeneous Poisson Process with rate $\frac{1}{3}$.

(a) $P(N(5) > N(3)) =$?
(b) $P(\{N(4) = 1\} \cap \{N(5) = 3\}) =$?
(c) $E(N(5)|N(3) = 2) =$?
(d) $E(t_6|N(3) = 4) =$?

Solution

(a) $P(N(5) > N(3)) = P(N(2) > 0)$ holds by *stationary* increments. Hence,

$$P(N(2) > 0) = 1 - P(N(2) = 0) = 1 - \frac{e^{-\left(\frac{1}{3}\right)(2)}\left(\left(\frac{1}{3}\right)(2)\right)^0}{0!} = 1 - e^{-\frac{2}{3}} = 0.4866$$

(b) We consider Fig. 9.6 for the solution.

$$P(\{N(4) = 1\} \cap \{N(5) = 3\})$$
$$= P(\{N(4) = 1\} \cap \{N(5 - 4) = 3 - 1\})$$
$$= P(\{N(4) = 1\} \cap \{N(1) = 2\})$$

by *stationary* increments, and

$$P(\{N(4) = 1\} \cap \{N(1) = 2\}) = P(N(4) = 1)P(N(1) = 2)$$

by *independent* increments. Finally,

$$P(N(4) = 1)P(N(1) = 2) = \left(\frac{e^{-\left(\frac{1}{3}\right)(4)}\left(\left(\frac{1}{3}\right)(4)\right)^1}{1!}\right)\left(\frac{e^{-\left(\frac{1}{3}\right)(1)}\left(\left(\frac{1}{3}\right)(1)\right)^2}{2!}\right) = \frac{2}{27}e^{-5/3} = 0.014$$

Interval length: 2

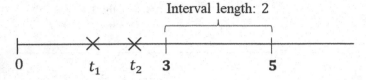

Fig. 9.7 Schematic representation of problem 2(c)

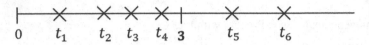

Fig. 9.8 Schematic representation of problem 2(d)

(c)   We consider Fig. 9.7 for the solution.

$E(N(5)|N(3) = 2) = 2 + E[N(5 - 3)] = 2 + E[N(2)]$ by *stationary* increments. Finally,

$$2 + E[N(2)] = 2 + \left(\frac{1}{3}\right)2 = \frac{8}{3} = 2.6667$$

(d)   We consider Fig. 9.8 for the solution.

$E(t_6|N(3) = 4) = 3 + E[t_2]$ by *stationary* increments. Finally,

$$3 + E[t_2] = 3 + \frac{2}{\left(\frac{1}{3}\right)} = 9$$

3.   Assume a two-server system. The service times for server 1 and server 2 are *independent* and *exponentially* distributed with rates λ_1 and λ_2, respectively. A customer can be served by any one of the two servers. A new customer enters the system, finds two servers busy, and no one waiting in queue. What is the expected total amount of time the new customer spends in the system?[2]

Solution
Let

T: Total amount of time the new customer spends in the system, where

$$T = S + W$$

S: Service time of the new customer

[2]This question is borrowed from Introduction to Probability Models, 10th Edition by Sheldon Ross.

W: Waiting time of the new customer for one of the servers to be free and let
X_1: Remaining service time of server 1
X_2: Remaining service time of server 2.

Hence, the expected total amount of time the new customer spends in the system will be

$$
\begin{aligned}
E[T] = E[S + W] &= E(S + W | X_1 < X_2) P(X_1 < X_2) \\
&\quad + E(S + W | X_2 < X_1) P(X_2 < X_1) \\
&= (E(S | X_1 < X_2) + E(W | X_1 < X_2)) P(X_1 < X_2) \\
&\quad + (E(S | X_2 < X_1) + E(W | X_2 < X_1)) P(X_2 < X_1) \\
&= \left(\frac{1}{\lambda_1} + \frac{1}{\lambda_1 + \lambda_2} \right) \left(\frac{\lambda_1}{\lambda_1 + \lambda_2} \right) + \left(\frac{1}{\lambda_2} + \frac{1}{\lambda_1 + \lambda_2} \right) \left(\frac{\lambda_2}{\lambda_1 + \lambda_2} \right) \\
&= \frac{3}{\lambda_1 + \lambda_2}
\end{aligned}
$$

Remark 4 Since an exponential random variable has *memoryless* property, the remaining service times are also exponentially distributed with the same parameter. For both cases, $X_1 < X_2$ and $X_2 < X_1$, $W = \min(X_1, X_2)$ is the remaining time until server 1 or server 2 finishes the service, where $W \sim \exp(\lambda_1 + \lambda_2)$.

4. Assume a two-server system. Service times at server 1 and server 2 are *independent* and *exponentially* distributed with rates 4 and 5 per hour, respectively. Every customer has to visit server 1 first, then server 2, and then should depart. When you enter the system, you find server 1 free, one customer (customer A) being served by server 2, and one customer (customer B) waiting in queue to be served by server 2. Find the probability that customer B is still in the system when you finish your service at server 1.[3]

Solution
Let

X_Y^1 : Your service time at server 1
X_A^2 : Remaining service time of customer A at server 2
X_B^2 : Service time of customer B at server 2

When you finish your service at server 1, there will be two *mutually exclusive* cases given that customer B is still in the system.

Case 1 You finish your service at server 1, but customer A does not finish the service at server 2, and customer B is still waiting to be served by server 2. Figure 9.9 illustrates this case.

Case 2 You finish your service at server 1, customer A has already finished the service at server 2, left the system, but customer B is still being served at server 2. Figure 9.10 illustrates this case.

---

[3]This question is adapted from Introduction to Probability Models, 10th Edition by Sheldon Ross

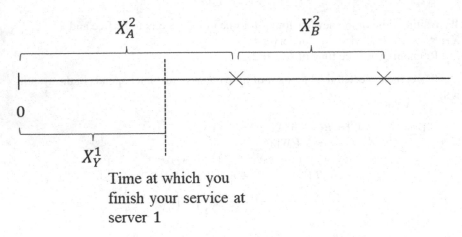

Time at which you
finish your service at
server 1

Fig. 9.9 Case 1 for problem 4

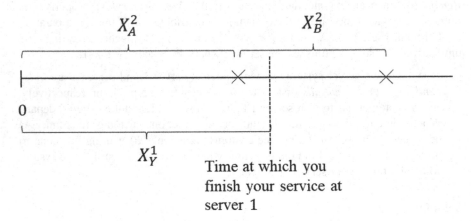

Time at which you
finish your service at
server 1

Fig. 9.10 Case 2 for problem 4

Let

P_B: Probability that customer B is still in the system when you finish your service at server 1.

Accordingly,

$$P_B = P\left(B \text{ is in system}\,\middle|\, X_Y^1 < X_A^2\right)P\left(X_Y^1 < X_A^2\right)$$
$$+ P\left(B \text{ is in system}\,\middle|\, X_Y^1 > X_A^2\right)P\left(X_Y^1 > X_A^2\right)$$
$$= 1\left(\frac{4}{(4+5)}\right) + \left(\frac{4}{(4+5)}\right)\left(\frac{5}{(4+5)}\right) = \frac{4}{9} + \left(\frac{4}{9}\right)\left(\frac{5}{9}\right) = \frac{56}{81}$$

Remark 5 $P\left(B \text{ is in system} \mid X_Y^1 < X_A^2\right) = 1$ follows since this is *Case 1*, in which customer A does not finish his service at server 2, and customer B is waiting in queue and in the system with probability 1. $P\left(B \text{ is in system } |X_Y^1 > X_A^2\right) = \frac{4}{9}$ follows since this is *Case* 2, in which the remaining service time of X_Y^1 after customer A finishes his service is still exponentially distributed with the same rate by *memoryless property*, thus, $P\left(B \text{ is in system } |X_Y^1 > X_A^2\right) = P(X_Y^1 < X_B^2) = \frac{4}{(4+5)} = \frac{4}{9}$ will hold.

5. Assume a *renewal process* with interarrival times normally distributed with mean 5 min and variance 20. What is the expected value and the variance of the time until the 7th renewal? Assume also that the interarrival times are Poisson distributed with mean 3 min. What is the expected value and the variance of the time until the 10th renewal?

Solution

Recall that t_n is the nth arrival time such that

$$t_n = X_1 + X_2 + X_3 + \cdots + X_n$$

Since

$$X_i \sim N(5, 20)\, i = 1, 2, 3, \ldots, n, \ldots$$

where X_i's are *i.i.d.*,

$$t_n \sim N(5n, 20n)$$

follows by the sum of *independent normal* random variables as provided in Table 7.3 in Chap. 7. Accordingly,

$$t_7 \sim N(35, 140)$$

which means

$$E[t_7] = 35$$
$$Var(t_7) = 140$$

Analogously, in case of *Poisson* distributed interarrival times,

$$t_n \sim Poiss(3n)$$

follows by the sum of *independent Poisson* random variables as provided in Table 7.3 in Chap. 7. Accordingly,

$$t_{10} \sim Poiss(30)$$

which means

$$E[t_{10}] = 30$$
$$Var(t_{10}) = 30$$

6. Consider an emergency service of a hospital which starts the service at 9:00 a.m.
    Patients arrive to this emergency service according to a Poisson process with
    varying arrival rates. The arrival rate between 9:00 a.m. and 10:00 a.m. is 3 per
    hour, after 10:00 a.m., there is a linear increase in the arrival rates between 10:00
    a.m. and 12:00, and at 12:00 there is a peak arrival rate of 6 per hour, which
    remains constant between 12:00–1:00 p.m. After 1:00 p.m., there is also a linear
    increase in the arrival rates until 3:00 p.m., and at 3:00 p.m. the arrival rate is
    assumed to be 8. After 3:00 p.m., there is a linear decrease in the arrival rates
    until 5:00 p.m., and at 5:00 p.m. the arrival rate is 6. Assume that we consider
    Poisson process between 9:00 a.m. and 5:00 p.m.

    (a) Plot and write the intensity function of this nonhomogeneous Poisson pro-
         cess.
    (b) What is the probability that 3 patients will arrive between 11:30 a.m. and
         2:30 p.m. given that 5 patients arrive between 9:00 a.m. 10:00 a.m.?

Solution

(a) We can plot the graph in Fig. 9.11 for $\lambda(t)$ by assuming 9:00 a.m. as $t = 0$.

    We can write the intensity function $\lambda(t)$ as

Fig. 9.11 Intensity function $\lambda(t)$ for problem 6

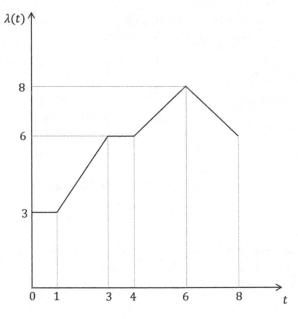

$$\lambda(t) = \begin{cases} 3 & 0 \le t \le 1 \\ \frac{3t+3}{2} & 1 \le t \le 3 \\ 6 & 3 \le t \le 4 \\ t+2 & 4 \le t \le 6 \\ -t+14 & 6 \le t \le 8 \end{cases}$$

(b)   The probability that 3 patients will arrive between 11:30 a.m. and 2:30 p.m. given that 5 patients arrive between 9:00 a.m. and 10:00 a.m. will be

$$P(N(5.5) - N(2.5) = 3 | N(1) = 5) = P(N(5.5) - N(2.5) = 3)$$

by *independent increments* since $(0, 1]$ and $(2.5, 5.5]$ are *disjoint* intervals.

Since $E[N(5.5) - N(2.5)] = 18.9375$ by finding the area under the intensity function $\lambda(t)$ for the interval $(2.5, 5.5]$,

$$P(N(5.5) - N(2.5) = 3) = \frac{e^{-18.9375}(18.9375)^3}{3!} = 0.000007$$

Chapter 10
Poisson Process

Abstract This chapter started with two formal definitions of a homogeneous Poisson process and a nonhomogeneous Poisson process, and the typical example of the arrivals of the customers. Some additional properties of a homogeneous Poisson process including partitioning a homogeneous Poisson process, superposition of a homogeneous Poisson process, determining the expected value of the number of events in an interval as a Poisson random variable approximation to a binomial random variable, the joint probability density function of the arrival times, and compound Poisson process were introduced. Some examples and problems including those related to the time of the earthquakes and the systems with different types of system failures were also presented.

While Chap. 9 provided the basic properties of a homogeneous and nonhomogeneous Poisson process, and the interarrival time distribution and the arrival time distribution for a homogeneous Poisson process; this chapter will provide two formal definitions for a homogeneous and nonhomogeneous Poisson process, additional properties of a homogeneous Poisson process including partitioning a homogeneous Poisson process, superposition of a homogeneous Poisson process, and compound Poisson process. The significance of the independent and stationary increments for a homogeneous Poisson process, and the importance of a nonhomogeneous Poisson process for representing the real-life random phenomena will be recalled with additional examples in this chapter. Some examples and problems will also be presented to illustrate the additional properties of a homogeneous Poisson process.

10.1 Homogeneous Versus Nonhomogeneous Poisson Process

Example 1 Suppose that a shop opens at 9:00 a.m. every day, and customers arrive to this shop according to a Poisson process.

© Springer Nature Switzerland AG 2019   149
E. Bas, *Basics of Probability and Stochastic Processes*,
https://doi.org/10.1007/978-3-030-32323-3_10

Definition 1 (*Homogeneous Poisson Process*) The renewal process

$$\{N(t); t \geq 0\}$$

is said to be a *homogeneous Poisson process* having rate λ, $\lambda > 0$ if

(i) $N(0) = 0$
(ii) The renewal process has *stationary* and *independent increments*.
(iii) $P(N(h) = 1) = \lambda h + o(h)$
(iv) $P(N(h) \geq 2) = o(h)$

where $o(h)$ is any function $f(h)$ with the following property:

$$\lim_{h \to 0} \frac{f(h)}{h} = 0$$

Definition 2 (*Homogeneous Poisson Process*) The renewal process

$$\{N(t); t \geq 0\}$$

is said to be a *homogeneous Poisson process* having rate λ, $\lambda > 0$ if

(i) $N(0) = 0$
(ii) The renewal process has *independent increments*.
(iii) The number of events in any interval is *Poisson* distributed such that the probability that k events will occur in the time interval $(s, s + t]$ will be

$$P(N(s + t) - N(s) = k) = P(N(t) = k) = \frac{e^{-\lambda t}(\lambda t)^k}{k!}$$

and

$$E[N(s + t) - N(s)] = E[N(t)] = \lambda t$$

$$Var(N(s + t) - N(s)) = Var(N(t)) = \lambda t$$

Definition 3 (*Nonhomogeneous Poisson Process*) The counting process

$$\{N(t); t \geq 0\}$$

is said to be a *nonhomogeneous Poisson process* having intensity function $\lambda(t), t \geq 0$ if

(i) $N(0) = 0$
(ii) The renewal process has *nonstationary* and *independent increments*.
(iii) $P(N(t + h) - N(t) = 1) = \lambda(t)h + o(h)$

(iv) $P(N(t + h) - N(t) \geq 2) = o(h)$

where $o(h)$ is any function $f(h)$ with the following property:

$$\lim_{h \to 0} \frac{f(h)}{h} = 0$$

Definition 4 (*Nonhomogeneous Poisson Process*) The counting process

$$\{N(t); t \geq 0\}$$

is said to be a *nonhomogeneous Poisson process* having intensity function $\lambda(t), t \geq 0$ if

(i) $N(0) = 0$
(ii) The renewal process has *independent increments*.
(iii) The number of events in any interval is *Poisson* distributed such that the probability that k events will occur in the time interval $(s, s + t]$ will be

$$P(N(s + t) - N(s) = k) = \frac{e^{-\int_s^{s+t} \lambda(y)dy} \left(\int_s^{s+t} \lambda(y)dy \right)^k}{k!}$$

and

$$E[N(s + t) - N(s)] = \int_s^{s+t} \lambda(y)dy$$

$$Var(N(s + t) - N(s)) = \int_s^{s+t} \lambda(y)dy$$

Remark 1 It can be shown that Definition 1 and Definition 2 are equivalent to each other, and Definition 3 and Definition 4 are equivalent to each other.

Example 1 (revisited 1) Assume that customers arrive according to a *homogeneous* Poisson process with rate 3 per hour. By considering 9:00 a.m. as $t = 0$, the probability that *at most* 4 customers will arrive between 2:00 p.m. and 4:00 p.m. given that 3 customers arrive between 10:00 a.m. and 1:00 p.m. will be

$$P(N(7) - N(5) \leq 4 \mid N(4) - N(1) = 3)$$
$$= P(N(7) - N(5) \leq 4)$$

by *independent increments* since the time intervals $(1, 4]$ and $(5, 7]$ are *disjoint*, and

$$P(N(7) - N(5) \le 4) = P(N(2) \le 4)$$

by *stationary increments*. Finally,

$$P(N(2) \le 4) = \sum_{k=0}^{4} P(N(2) = k) = \sum_{k=0}^{4} \frac{e^{-(3)(2)}((3)(2))^k}{k!} = \sum_{k=0}^{4} \frac{e^{-6}(6)^k}{k!}$$

$$= e^{-6} + \left(\frac{6^1}{1!}\right)e^{-6} + \left(\frac{6^2}{2!}\right)e^{-6} + \left(\frac{6^3}{3!}\right)e^{-6} + \left(\frac{6^4}{4!}\right)e^{-6} = 0.2851$$

and

$$E[N(7) - N(5)] = E[N(2)] = (3)(2) = 6$$

$$Var(N(7) - N(5)) = Var(N(2)) = (3)(2) = 6$$

Example 1 (revisited 2) Assume now that customers arrive according to a *nonhomogeneous* Poisson process with intensity function

$$\lambda(t) = \begin{cases} 2t + 1 & 0 \le t \le 6 \\ 13 & t \ge 6 \end{cases}$$

Hence, the probability that *at most* 4 customers will arrive between 2:00 p.m. and 4:00 p.m. given that 3 customers arrive between 10:00 a.m. and 1:00 p.m. will be

$$P(N(7) - N(5) \le 4 \mid N(4) - N(1) = 3)$$
$$= P(N(7) - N(5) \le 4)$$

by *independent increments* since the time intervals $(1, 4]$ and $(5, 7]$ are *disjoint*.

To compute $P(N(7) - N(5) \le 4)$, we can start by computing $E[N(7) - N(5)]$ such that

$$E[N(7) - N(5)] = \int_{5}^{6} (2t + 1)dt + \int_{6}^{7} 13dt = 25$$

Accordingly,

$$P(N(7) - N(5) \le 4) = \sum_{k=0}^{4} \frac{e^{-25}(25)^k}{k!}$$

$$= e^{-25} + \left(\frac{25^1}{1!}\right)e^{-25} + \left(\frac{25^2}{2!}\right)e^{-25} + \left(\frac{25^3}{3!}\right)e^{-25}$$

$$+ \left(\frac{25^4}{4!}\right) e^{-25} = 2.67 \times 10^{-7}$$

10.2 Additional Properties of a Homogeneous Poisson Process

Definition 5 (*Partitioning a homogeneous Poisson process, thinning a homogeneous Poisson process*) Let $\{N(t); t \geq 0\}$ be a *homogeneous* Poisson Process with rate λ. If each event can be classified *independently* as Type 1 or Type 2 event with probabilities p and $1 - p$, respectively, then

$$\{N_1(t); t \geq 0\} \text{ and } \{N_2(t); t \geq 0\}$$

can be defined as two *independent homogeneous* Poisson processes with rates λp and $\lambda(1 - p)$, respectively, where

$$N_1(t): \text{Number of Type 1 events in } (0, t] \text{ with } E[N_1(t)] = \lambda p t$$

$$N_2(t): \text{Number of Type 2 events in } (0, t] \text{ with } E[N_2(t)] = \lambda(1 - p)t$$

Note that

$$N(t) = N_1(t) + N_2(t) \quad \text{for } t \geq 0$$

Example 1 (revisited 3) Assume now that customers arrive according to a Poisson process with rate 3 per hour, and assume also that each customer has car with probability $\frac{2}{3}$. If we define

Type 1: Customer with car
Type 2: Customer with no car

then the probability that 4 customers *with* car will arrive between 2:00 p.m. and 4:00 p.m. given that 3 customers *with* car arrive between 10:00 a.m. and 1:00 p.m. will be

$$P(N_1(7) - N_1(5) = 4 \mid N_1(4) - N_1(1) = 3) = P(N_1(7) - N_1(5) = 4)$$

by *independent increments*, and

$$P(N_1(7) - N_1(5) = 4) = P(N_1(2) = 4)$$

by *stationary increments*. Finally,

$$0 \quad \frac{t}{k} \quad \frac{2t}{k} \quad \frac{3t}{k} \qquad\qquad \cdots\cdots\cdots \qquad\qquad t = \frac{kt}{k}$$

Fig. 10.1 Partitioning the time interval $(0, t]$ into k *equal* subintervals

$$P(N_1(2) = 4) = \frac{e^{-(3)\left(\frac{2}{3}\right)(2)}\left((3)\left(\frac{2}{3}\right)(2)\right)^4}{4!} = \frac{e^{-4}(4)^4}{4!} = 0.1954$$

Definition 6 (*Superposition of a homogeneous Poisson process*) Let $\{N_1(t); t \geq 0\}$ and $\{N_2(t); t \geq 0\}$ be *any* two *independent homogeneous* Poisson processes with rates λ_1 and λ_2, respectively. These Poisson processes can be superposed such that

$$\{N(t); t \geq 0\}$$

is a *homogeneous* Poisson process with rate $\lambda = \lambda_1 + \lambda_2$, where

$$N(t) = N_1(t) + N_2(t) \quad \text{for } t \geq 0$$

Remark 2 Definition 5 and Definition 6 can be extended to any number of homogeneous Poisson processes $r, r \geq 2$ with rates $\lambda_1, \lambda_2, \ldots, \lambda_r$ such that $N(t) = N_1(t) + N_2(t) + \cdots + N_r(t)$, and $\lambda_1 + \lambda_2 + \cdots + \lambda_r = \lambda$.

Example 1 (revisited 4) Assume that customers *with* car arrive according to a Poisson process with rate 2 per hour, and customers *with no* car arrive according to a Poisson process with rate 1 per hour. Accordingly, the probability that 5 customers will arrive between 10:00 a.m. and 2:00 p.m. will be

$$P(N(5) - N(1) = 5) = P(N(4) = 5) = \frac{e^{-(2+1)4}((2+1)(4))^5}{5!}$$

$$= \frac{e^{-12}(12)^5}{5!} = 0.0127$$

Definition 7 (*Determining $E[N(t)]$ as a Poisson random variable approximation to a binomial random variable for a homogeneous Poisson process*)[1]

Let the time interval $(0, t]$ be partitioned into k *equal* subintervals as illustrated in Fig. 10.1.

It can be shown that as $k \to \infty$, in each subinterval, there can be 1 or 0 event, thus, each subinterval can be regarded as a trial as illustrated in Fig. 10.2.

Recall that

$$N(t) : \text{ Number of events in } (0, t]$$

---

[1]Borrowed from Stochastic Processes by Sheldon Ross.

Fig. 10.2 Consideration of
each subinterval as a trial

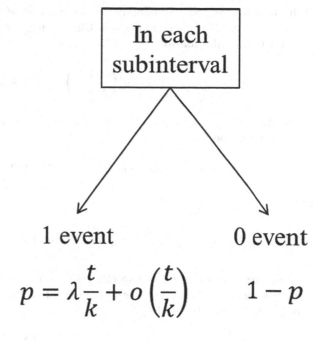

$$p = \lambda\frac{t}{k} + o\left(\frac{t}{k}\right) \qquad 1 - p$$

Since there are k subintervals in $(0, t]$, i.e. k trials, each of which with probability p for 1 event,

$$N(t) \sim Bin(k, p)$$

and by Poisson random variable *approximation* to binomial random variable, it follows that

$$E[N(t)] = \lim_{k\to\infty} kp = \lim_{k\to\infty}\left(k\left(\lambda\frac{t}{k} + o\left(\frac{t}{k}\right)\right)\right)$$

$$= \lambda t + \lim_{k\to\infty}\left(ko\left(\frac{t}{k}\right)\right)$$

$$= \lambda t + \lim_{k\to\infty}\left(\frac{o\left(\frac{t}{k}\right)}{\frac{1}{k}}\right)$$

$$= \lambda t + \lim_{k\to\infty} t\left(\frac{o\left(\frac{t}{k}\right)}{\frac{t}{k}}\right) = \lambda t + t\lim_{k\to\infty}\left(\frac{o\left(\frac{t}{k}\right)}{\frac{t}{k}}\right) = \lambda t$$

Remark 3 Since $P(N(h) = 1) = \lambda h + o(h)$ by Definition 1, it follows that $P\left(N\left(\frac{t}{k}\right) = 1\right) = p = \lambda\frac{t}{k} + o\left(\frac{t}{k}\right)$. Additionally, $\lim_{k\to\infty}\left(\frac{o\left(\frac{t}{k}\right)}{\frac{t}{k}}\right) = 0$ due to the definition of an $o(h)$ function provided in Definition 1 by noting that $\lim_{k\to\infty}\frac{t}{k} = 0$.

Definition 8 (*The joint probability density function of $t_1, t_2, \ldots, t_n$ given that $N(t) = n$*) Recall from Example 4 in Chap. 7 that the joint probability density function of the order statistics of $X_1, X_2, \ldots, X_n$ is

$$f_{X_{(1)}, X_{(2)}, \ldots, X_{(n)}}(x_1, x_2, \ldots, x_n) = n! \left(\frac{1}{\beta - \alpha} \right)^n \quad \text{for } \alpha < x_1 < x_2 < \cdots < x_n < \beta$$

where $X_i \sim Uniform(\alpha, \beta), i = 1, 2, \ldots, n$ are *independent* random variables.
    As a special case, if

$$X_i \sim Uniform(0, t), \quad \text{for } i = 1, 2, \ldots, n$$

then

$$f_{X_{(1)}, X_{(2)}, \ldots, X_{(n)}}(x_1, x_2, \ldots, x_n) = n! \left(\frac{1}{t} \right)^n \quad \text{for } 0 < x_1 < x_2 < \cdots < x_n < t$$

It can be shown that for the special case $N(t) = n$, where $\{N(t); t \geq 0\}$ is a *homogeneous* Poisson process, the joint probability density function of the arrival times $t_1, t_2, \ldots, t_n$ will also be

$$f(t_1, t_2, \ldots, t_n) = n! \left(\frac{1}{t} \right)^n$$

Remark 4 For the very special case $N(t) = 1$,

$$f(t_1) = \frac{1}{t}$$

and

$$P(t_1 < s) = \frac{s}{t}$$

Example 1 (revisited 5) Assume again that customers arrive according to a Poisson process with rate 3 per hour, and assume also that $N(5) = 6$. Hence,

$$f(t_1, t_2, \ldots, t_6) = 6! \left(\frac{1}{5} \right)^6 = 0.0461$$

Definition 9 (*Compound Poisson process, batch Poisson process*)

$$\{X(t); t \geq 0\}$$

is a special continuous-time stochastic process called as a *compound Poisson process*, where

$$X(t) = \sum_{i=1}^{N(t)} X_i$$

Here, X_i's are *i.i.d.* random variables, and $\{N(t); t \geq 0\}$ is a *homogeneous* Poisson process with rate λ and *independent* from X_i's. It can be shown that

$$E[X(t)] = \lambda t E[X_i]$$

$$Var(X(t)) = \lambda t E[X_i^2]$$

Remark 5 Each $X(t)$ random variable is called as a *compound Poisson random variable*.

Example 1 (revisited 6) Assume that customers arrive according to a Poisson process with rate 3 per hour, and assume also that each customer spends *exponentially* distributed amount of money with mean \$35. As an example, the total amount of money customers spend between 11:00 a.m. and 3:00 p.m. will be $X(4)$ due to the *stationary increments* of a Poisson process and

$$E[X(4)] = (3)(4)E[X_i]$$

by defining

$$X_i : \text{ Amount of money customer } i \text{ spends}$$

and by noting that

$$E[X_i] = 35 \quad \forall i$$

$$\lambda = \frac{1}{E[X_i]} = \frac{1}{35}$$

Finally,

$$E[X(4)] = (3)(4)E[X_i] = (3)(4)(35) = \$420$$

The variance of the total amount of money customers spend between 11:00 a.m. and 3:00 p.m. will be

$$Var(X(4)) = (3)(4)E[X_i^2]$$

where

$$E[X_i^2] = Var(X_i) + (E[X_i])^2$$
$$= \frac{1}{\left(\frac{1}{35}\right)^2} + (35)^2 = 2{,}450$$

Finally,

$$Var(X(4)) = (3)(4)E[X_i^2] = (3)(4)(2{,}450) = 29{,}400$$

Problems

1. It is assumed that earthquakes occur according to a Poisson process at a rate of $\frac{1}{2}$ per year. Each time an earthquake occurs, the probability of its Richter scale to be *higher* than 6 is 0.3. Given that 2 earthquakes occur in the following 15 years with a Richter scale *higher* than 6, what is the probability that 3 earthquakes will occur with a Richter scale *lower* than 6 between the years $2{,}050 - 2{,}055$?

Solution

We can assume each earthquake as

Type 1: Earthquake with a Richter scale *higher* than 6, with probability 0.3
Type 2: Earthquake with a Richter scale *lower* than 6, with probability 0.7

Accordingly, $\{N_1(t); t \geq 0\}$ and $\{N_2(t); t \geq 0\}$ will be the Poisson processes for Type 1 and Type 2 earthquakes, respectively, and

$$P(N_2(5) = 3 \mid N_1(15) = 2) = P(N_2(5) = 3)$$

due to the *independence* of $\{N_1(t); t \geq 0\}$ and $\{N_2(t); t \geq 0\}$, and due to the *stationary increments* of a homogeneous Poisson process. Finally,

$$P(N_2(5) = 3) = \frac{e^{-\left(\frac{1}{2}\right)(0.7)(5)}\left(\left(\frac{1}{2}\right)(0.7)(5)\right)^3}{3!} = \frac{e^{-1.75}(1.75)^3}{3!} = 0.1552$$

2. Assume for a system that system failures occur according to a homogeneous Poisson process $\{N(t); t \geq 0\}$ with rate λ, and let

$$t_i : \text{Time of the } i\text{th system failure}$$

Let also s and t be the fixed times such that $s \leq t$. Find $E\left(\sum_{i=1}^{N(t)} (s - t_i)^2\right)$.

Solution

By conditioning on $N(t) = n$,

$$E\left(\sum_{i=1}^{N(t)}(s - t_i)^2 | N(t) = n\right) = E\left(\sum_{i=1}^{N(t)}(s^2 - 2st_i + t_i^2) | N(t) = n\right)$$

Given that $N(t) = n$,

$$t_i \sim Uniform(0, t)$$

from Definition 8 in this chapter, and

$$E[t_i] = \frac{0+t}{2} = \frac{t}{2}$$

$$Var(t_i) = \frac{(t-0)^2}{12} = \frac{t^2}{12}$$

$$E[t_i^2] = Var(t_i) + (E[t_i])^2 = \frac{t^2}{12} + \left(\frac{t}{2}\right)^2 = \frac{t^2}{3}$$

Due to the identity $E[X] = E[E(X | Y)]$ provided in Table 7.6 in Chap. 7, it follows that

$$E\left(E\left(\sum_{i=1}^{N(t)}(s^2 - 2st_i + t_i^2) | N(t) = n\right)\right) = E\left(\sum_{i=1}^{N(t)}(s^2 - 2st_i + t_i^2)\right)$$

$$= E[N(t)]s^2 - 2s\,E[N(t)]E[t_i] + E[N(t)]E[t_i^2]$$

$$= \lambda t s^2 - 2s\lambda t\left(\frac{t}{2}\right) + \lambda t\left(\frac{t^2}{3}\right)$$

$$= \lambda t s^2 - \lambda s t^2 + \frac{\lambda t^3}{3}$$

3. A system's failures occur according to a Poisson process with rate $\frac{1}{4}$ per year. Assume that whenever a system failure occurs, it is Type 1 failure with probability 0.3, Type 2 failure with probability 0.2, and Type 3 failure with probability 0.5. *Approximate* the probability that there will be *at least* 20 Type 1 and Type 2 system failures by year $t = 100$.

Solution

Let
$\{N_1(t); t \geq 0\}$ be a Poisson process with rate $\left(\frac{1}{4}\right)0.3 = \frac{3}{40}$, where
$N_1(t)$ is the number of Type 1 system failures in $(0, t]$,

$\{N_2(t); t \geq 0\}$ be a Poisson process with rate $\left(\frac{1}{4}\right)0.2 = \frac{2}{40}$, where $N_2(t)$ is the number of Type 2 system failures in $(0, t]$,
$\{N_3(t); t \geq 0\}$ be a Poisson process with rate $\left(\frac{1}{4}\right)0.5 = \frac{5}{40}$, where $N_3(t)$ is the number of Type 3 system failures in $(0, t]$.

Hence, the *approximated* probability that there will be *at least* 20 Type 1 and Type 2 system failures by year $t = 100$ will be

$$P(N_1(100) + N_2(100) \geq 19.5)$$

by considering the *continuity correction*. Note that for Type 1 system failures,

$$E[N_1(100)] = \left(\frac{3}{40}\right)100 = 7.5, \ Var(N_1(100)) = 7.5$$

and for Type 2 system failures,

$$E[N_2(100)] = \left(\frac{2}{40}\right)100 = 5, \ Var(N_2(100)) = 5$$

Accordingly,

$$P\left(\frac{N_1(100) + N_2(100) - (E[N_1(100)] + E[N_2(100)])}{\sqrt{Var(N_1(100)) + Var(N_2(100))}} \geq \frac{19.5 - (7.5 + 5)}{\sqrt{(7.5 + 5)}}\right)$$

$$= P\left(Z \geq \frac{7}{\sqrt{12.5}}\right) = P(Z \geq 1.9799) = 1 - \Phi(1.9799)$$

$$\cong 1 - 0.9761 = 0.0239$$

4. In a hospital, for a given day, pregnant women give birth according to a homogeneous Poisson process with rate $\frac{1}{3}$ per hour. Each woman gives birth at a time to 1 baby with probability 0.7, gives birth to 2 babies with probability 0.2, and gives birth to 3 babies with probability 0.1 *independently* from each other. What is the expected value and the variance of the number of the newborn babies in a day in this hospital?

Solution

Let

$$\{X(t); t \geq 0\}$$

be a *compound Poisson process*, where

$$X(t) = \sum_{i=1}^{N(t)} X_i$$

$X(t)$: Number of the newborn babies in $(0, t]$

$N(t)$: Number of the pregnant women who give birth in $(0, t]$

X_i: Number of the newborn babies of the ith pregnant woman who gives birth in $(0, t]$

Accordingly,

$X(24)$: Number of the newborn babies in a day

By recalling that

$$E[X(t)] = \lambda t E[X_i]$$

$$Var(X(t)) = \lambda t E[X_i^2]$$

it follows that

$$E[X(24)] = \left(\frac{1}{3}\right)24E[X_i] = \left(\frac{1}{3}\right)24((1)(0.7) + (2)(0.2) + (3)(0.1)) = 11.2$$

$$Var(X(24)) = \left(\frac{1}{3}\right)24E[X_i^2] = \left(\frac{1}{3}\right)24\left((1^2)(0.7) + (2^2)(0.2) + (3^2)(0.1)\right) = 19.2$$

Chapter 11
Renewal Process

Abstract The first part of this chapter has been devoted to the basic terms related to a renewal process including the renewal function and renewal equation. The second part provided important theorems such as limit theorem, elementary renewal theorem, and renewal reward process, and some other properties such as the excess, age, and spread of a renewal process, inspection paradox, and central limit theorem for a renewal process. In the third part of the chapter, the alternating renewal process has been introduced as an example of a regenerative process. The concepts were illustrated by using some examples and problems including the ones related to the departures of the buses, and the system failures.

Chapter 9 provided the basic definition of a renewal process with one example and one problem, but did not provide any additional properties. As an important type of a counting process with *i.i.d.* interarrival times, this chapter will provide the basic terms related to a renewal process such as the renewal function and renewal equation, important theorems including limit theorem, elementary renewal theorem, renewal reward process, and central limit theorem for a renewal process, a well-known paradox called as inspection paradox, and alternating renewal process. It should be recalled that homogeneous *Poisson process* provided in Chap. 10 is a special type of a renewal process with *exponentially* distributed interarrival times. Thus, the learnings from this chapter can be extended to homogeneous Poisson process, and this extension opportunity will be illustrated in one of the problems of this chapter. There will be also other representative examples and problems related to the properties of a renewal process.

11.1 Basic Terms

Example 1 A system's lifetime is *exponentially* distributed with rate $\frac{1}{40}$ per hour. After breakdown, it is repaired for *uniformly* distributed time defined over $(0, 1)$

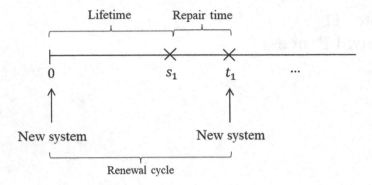

Fig. 11.1 System renewal problem

hour, and after repair the system becomes like new again. This process continues infinitely with *independent and identical* distributions, and can be illustrated as in Fig. 11.1.

Definition 1 (*Renewal process*) A renewal process is a *counting process* $\{N(t); t \geq 0\}$ in which the stochastic process of the *interarrival* times is *i.i.d.*, i.e., $\{X_n; n \geq 1\}$ are *i.i.d.*, where $E[X_n] = E[X] = \mu, n \geq 1$.

Remark 1 The interarrival time for a renewal process can also be called as the *length of a renewal cycle*.

Definition 2 (*Probability mass function of $N(t)$ for a renewal process*) For a *counting process* $\{N(t); t \geq 0\}$, note that $N(t) \geq n \leftrightarrow t_n \leq t$ holds, and can be illustrated as in Fig. 11.2.
    Analogously,

$$N(t) \geq n + 1 \leftrightarrow t_{n+1} \leq t$$

holds. Hence,

$$
\begin{aligned}
P(N(t) = n) &= P(N(t) \geq n) - P(N(t) \geq n + 1) \\
&= P(t_n \leq t) - P(t_{n+1} \leq t) \\
&= P(X_1 + X_2 + \cdots + X_n \leq t) - P(X_1 + X_2 + \cdots + X_{n+1} \leq t)
\end{aligned}
$$

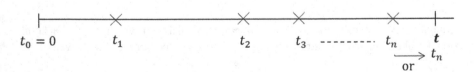

Fig. 11.2 Schematic representation of $N(t) \geq n \leftrightarrow t_n \leq t$

Special case
For a renewal process $\{N(t); t \geq 0\}$,

$$P(N(t) = n) = P(X_1 + X_2 + \ldots + X_n \leq t) - P(X_1 + X_2 + \ldots + X_{n+1} \leq t)$$
$$= F_n(t) - F_{n+1}(t)$$

since $\{X_n; n \geq 1\}$ are *i.i.d.*, where $F_n(t)$ and $F_{n+1}(t)$ are called as the n-fold and $n+1-$ fold convolution of the cumulative distribution functions F of the interarrival times.

Definition 3 (*Renewal function*) For a renewal process $\{N(t); t \geq 0\}$,

$$m(t) = E[N(t)]$$

is called as a *renewal function*. It can be shown that

$$m(t) = E[N(t)] = \sum_{n=1}^{\infty} F_n(t)$$

where $F_n(t)$ is the n-fold convolution of the cumulative distribution functions F of the interarrival times.

Definition 4 (*Renewal equation*) For a renewal process $\{N(t); t \geq 0\}$,

$$m(t) = F(t) + \int_0^t m(t - x) f(x) dx$$

is called as a *renewal equation*, where F and f are the cumulative distribution functions and the probability density functions of the interarrival times, respectively, and $m(t - x) = E[N(t - x)]$.

11.2 Limit Theorem, Elementary Renewal Theorem, Renewal Reward Process

Definition 5 (*Limit theorem vs. elementary renewal theorem vs. renewal reward process*) Table 11.1 provides the definitions of the limit theorem, elementary renewal theorem, and renewal reward process.

Remark 2 Limit theorem and *elementary renewal theorem* are related to the *long-run average rate* of an event in a renewal process, while *renewal reward process* is related to the *long-run average reward or cost* in a renewal process. Note that the consideration can be *reward* or *cost* in a renewal reward process.

Table 11.1 Limit theorem versus elementary renewal theorem versus renewal reward process

Limit theorem	Elementary renewal theorem (ERT)	Renewal reward process
$\frac{N(t)}{t} \to \frac{1}{E[X]}$ as $t \to \infty$ $E[X]$: Expected value of an interarrival time (*length of a renewal cycle*) of a renewal process	$\frac{m(t)}{t} \to \frac{1}{E[X]}$ as $t \to \infty$ $m(t)$ is the renewal function.	$\frac{R(t)}{t} \to \frac{E[R]}{E[X]}$ as $t \to \infty$ $\frac{E[R(t)]}{t} \to \frac{E[R]}{E[X]}$ as $t \to \infty$ $R(t) = \sum\limits_{n=1}^{N(t)} R_n$ is the total rewards earned in $(0, t]$. R_n: Reward earned at the time of the n th renewal such that $\{R_n; n \geq 1\}$ are *i.i.d.*, $E[R_n] = E[R]$ for $n \geq 1$. $\{N(t); t \geq 0\}$ is a renewal process.

Example 1 (revisited 1) Recall that the system's lifetime is *exponentially* distributed with rate $\frac{1}{40}$ per hour, and the repair time is *uniformly* distributed over $(0, 1)$ hour. Hence, the long-run average rate at which the system is renewed will be

$$\frac{N(t)}{t} \to \frac{1}{E[X]} \text{ as } t \to \infty$$

from *limit theorem*, where

$N(t)$: Number of system renewals in $(0, t]$

and

$$E[X] = E\left[\text{length of a renewal cycle}\right] = E[\text{lifetime}] + E\left[\text{repair time}\right]$$
$$= \frac{1}{\left(\frac{1}{40}\right)} + \frac{(0+1)}{2} = 40 + \frac{1}{2} = 40.5 \text{ h}$$

Finally,

$$\frac{N(t)}{t} \to \frac{1}{40.5} \text{ as } t \to \infty$$

In words, in the long-run, on average, the system is renewed every 40.5 h.

Example 1 (revisited 2) Each time the system breaks down, there is a repair cost that is *uniformly* distributed over $(100, 200)$ dollars. Hence, the *long-run average cost* will be

$$\frac{E[C]}{E[X]} = \frac{E\left[\text{cost of a renewal cycle}\right]}{E\left[\text{length of a renewal cycle}\right]}$$

by *renewal reward process*, where

$$E\left[\text{cost of a renewal cycle}\right] = \frac{(100 + 200)}{2} = \$150$$

and

$$E\left[\text{length of a renewal cycle}\right] = 40.5 \, \text{h}$$

from Example 1 (revisited 1). Finally,

$$\frac{E[C]}{E[X]} = \frac{150}{40.5} = \$3.7037 \text{ per hour}$$

Definition 6 (*Age vs. excess vs. spread of a renewal process*) We consider Fig. 11.3 for understanding the definitions given in Table 11.2

Remark 3 Note that $t_{N(t)}$ refers to the time of the last event in $(0, t]$. Analogously, $t_{N(t)+1}$ refers to the time of the first event after time t. The *average age* and *average excess* of a renewal process can be derived by using the definition of *renewal reward process*.

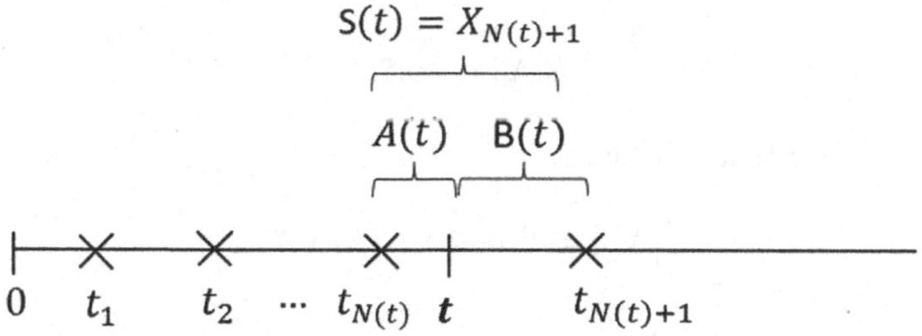

Fig. 11.3 Age, excess, and spread of a renewal process

Table 11.2 Basic formulae for the age, excess, and spread of a renewal process

Age of a renewal process (Backwards recurrence time) at time t	Excess of a renewal process (Forward recurrence time) at time t	Spread of a renewal process at time t
$A(t) = t - t_{N(t)}, \quad t \geq 0$	$B(t) = t_{N(t)+1} - t, \quad t \geq 0$	$S(t) = X_{N(t)+1}$ $= t_{N(t)+1} - t_{N(t)}$ $= A(t) + B(t) \quad t \geq 0$
Average age of a renewal process	**Average excess of a renewal process**	**Average spread of a renewal process**
$\dfrac{E[X^2]}{2E[X]}$ X: Interarrival time (*length of a renewal cycle*) of a renewal process	$\dfrac{E[X^2]}{2E[X]}$	$\dfrac{E[X^2]}{E[X]}$

Example 1 (revisited 3) If we consider the time *from* the *last* system renewal to $t = 40$, then the *age* of the renewal process at $t = 40$ will be

$$A(40) = 40 - t_{N(40)}$$

If we consider the time *from* $t = 40$ *until* the *next* system renewal, then the *excess* of the renewal process at $t = 40$ will be

$$B(40) = t_{N(40)+1} - 40$$

If we consider the time *between* the *last* system renewal until $t = 40$ and the *next* system renewal after $t = 40$, then the *spread* of the renewal process at $t = 40$ will be

$$S(40) = X_{N(40)+1} = t_{N(40)+1} - t_{N(40)} = A(40) + B(40)$$

Recall that the *average age* or the *average excess* of the renewal process is

$$\frac{E[X^2]}{2E[X]}$$

Note that

$$E[X] = 40.5$$

from Example 1 (revisited 1), and

$$E[X^2] = Var(X) + (E[X])^2$$

where

$$Var(X) = Var(\text{lifetime}) + Var(\text{repair time})$$

$$= \frac{1}{\left(\frac{1}{40}\right)^2} + \frac{(1-0)^2}{12} = 1{,}600 + \frac{1}{12} = 1{,}600.083$$

Accordingly,

$$E[X^2] = Var(X) + (E[X])^2 = 1{,}600.083 + (40.5)^2 = 3{,}240.33$$

and

$$\frac{E[X^2]}{2E[X]} = \frac{3{,}240.33}{2(40.5)} = 40.004\,\text{h}$$

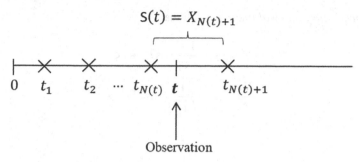

Fig. 11.4 Inspection paradox

Definition 7 (*Inspection paradox*) We consider a system which is replaced immediately with a new one in case of a failure. Let $\{N(t); t \geq 0\}$ be a renewal process, where $N(t)$ is the number of the failures/immediate replacements of this system in $(0, t]$.

We assume that we make an *observation* to this system at time t, and find that the system functions as illustrated in Fig. 11.4. It can be shown that

$$P(S(t) > x) \geq P(X > x)$$

holds, where X is the interarrival time between the consecutive failures/immediate replacements of this system *without observation*. This phenomenon is called as an *inspection paradox*. Inspection paradox means that the system observed to be functioning tends to have a larger or equal lifetime with respect to the system *without observation*.

Definition 8 (*Central limit theorem for a renewal process*) Let $\{N(t); t \geq 0\}$ be a renewal process. Accordingly,

$$\frac{N(t) - E[N(t)]}{\sqrt{Var(N(t))}} = \frac{N(t) - t/\mu}{\sqrt{t\sigma^2/\mu^3}} \to Z \sim N(0, 1) \quad \text{as } t \to \infty$$

where

$$\mu = E[X]$$

$$\sigma^2 = Var(X)$$

are the expected value and the variance of the interarrival times, respectively.

Remark 4 Note that

$$E[N(t)] = t/\mu$$

from *elementary renewal theorem* such that

$$\frac{m(t)}{t} \to \frac{1}{E[X]} \quad \text{as } t \to \infty$$

$$\frac{E[N(t)]}{t} \to \frac{1}{E[X]} \quad \text{as } t \to \infty$$

and

$$E[N(t)] \to \frac{t}{E[X]} \quad \text{as } t \to \infty$$

Finally,

$$E[N(t)] \to \frac{t}{\mu} \quad \text{as } t \to \infty$$

since $\mu = E[X]$.
Note also that

$$\frac{Var(N(t))}{t} \to \frac{\sigma^2}{\mu^3} \quad \text{as } t \to \infty$$

and

$$Var(N(t)) \to \frac{t\sigma^2}{\mu^3} \quad \text{as } t \to \infty$$

11.3 Regenerative Process

Recall the classical definition of a regenerative process from Definition 6 in Sect. 9.2. Recall also that a renewal process is special type of a regenerative process. Definition 9 provided in this section is an example of a regenerative process in the context of a renewal process.

Definition 9 (*Alternating renewal process*) We consider a special renewal process called as an *alternating renewal process* which continues *infinitely* as illustrated in Fig. 11.5.

In an alternating renewal process,
$\{Z_n, Y_n\}$ are *i.i.d.* for $n \geq 1$ with

Z_n: Time of *on-phase* in the nth renewal
Y_n: Time of *off-phase* in the nth renewal

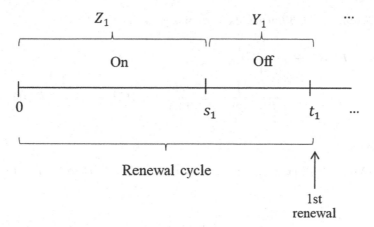

Fig. 11.5 Alternating renewal process

where

$$E[Z_n] = E[Z] \text{ for } \forall n, \quad E[Y_n] = E[Y] \text{ for } \forall n$$

It can be shown that

$$\lim_{t \to \infty} P_{on}(t) = \frac{E[Z]}{E[Z] + E[Y]}$$

$$\lim_{t \to \infty} P_{off}(t) = \frac{E[Y]}{E[Z] + E[Y]} = 1 - \lim_{t \to \infty} P_{on}(t)$$

where

$P_{on}(t)$: Probability that the system is *on* at time t
$P_{off}(t)$: Probability that the system is *off* at time t
$\lim_{t \to \infty} P_{on}(t)$: Limiting probability that the system is *on* (Long-run proportion of time that the system is *on*)
$\lim_{t \to \infty} P_{off}(t)$: Limiting probability that the system is *off* (Long-run proportion of time that the system is *off*)

Example 1 (revisited 4) The *limiting probability* that the system is *on* will be

$$\lim_{t \to \infty} P_{on}(t) = \frac{E[Z]}{E[Z] + E[Y]}$$

$$= \frac{\left(\frac{1}{40}\right)}{\left(\frac{1}{40}\right) + \left(\frac{0+1}{2}\right)} = \frac{40}{40.5} = 0.9877$$

and the *limiting probability* that the system is *off* will be

$$\lim_{t \to \infty} P_{off}(t) = \frac{E[Y]}{E[Z] + E[Y]}$$

$$= \frac{\left(\frac{0+1}{2}\right)}{\left(\frac{1}{40}\right) + \left(\frac{0+1}{2}\right)} = \frac{0.5}{40.5} = 0.0123 = 1 - \lim_{t \to \infty} P_{on}(t)$$

Problems

1. Let $\{N(t); t \geq 0\}$ be a renewal process where $\{X_n; n \geq 1\}$ are *i.i.d.* with

$$X_n \sim Poiss(\lambda), n \geq 1$$

Calculate $P(N(t) = n)$.

Solution
Recall that

$$t_n = X_1 + X_2 + \cdots + X_n$$

Since

$$X_n \sim Poiss(\lambda) \quad n \geq 1$$

are independent,

$$t_n \sim Poiss(n\lambda) \quad n \geq 1$$

by the sum of *independent Poisson* random variables as provided in Table 7.3 in Chap. 7. Hence,

$$P(t_n = k) = \frac{e^{-n\lambda}(n\lambda)^k}{k!}$$

Analogously,

$$t_{n+1} \sim Poiss((n+1)\lambda)$$

and

$$P(t_{n+1} = k) = \frac{e^{-(n+1)\lambda}((n+1)\lambda)^k}{k!}$$

will follow. By recalling Definition 2 in this chapter,

$$P(N(t) = n) = P(t_n \leq t) - P(t_{n+1} \leq t)$$

$$= \sum_{k=0}^{\lfloor t \rfloor} \frac{e^{-n\lambda}(n\lambda)^k}{k!} - \sum_{k=0}^{\lfloor t \rfloor} \frac{e^{-(n+1)\lambda}((n+1)\lambda)^k}{k!}$$

Remark 5 Note that $\lfloor . \rfloor$ is a *floor function* which yields the largest integer less than or equal to the value it contains. Since $t_n \sim Poiss(n\lambda)$ and $t_{n+1} \sim Poiss((n+1)\lambda)$ are *discrete* random variables, but t is allowed to take any positive real number, we use floor function to find t as an integer number.

2. Please show that $m(t) = E[N(t)] = \lambda t$ for a *homogeneous* Poisson process by using the renewal equation.

Solution
Recall the renewal equation as

$$m(t) = F(t) + \int_0^t m(t-x)f(x)dx$$

from Definition 4 in this chapter. Since $X_n \sim \exp(\lambda)$ $n \geq 1$ for a *homogeneous* Poisson process,

$$F(t) = 1 - e^{-\lambda t}$$

$$f(x) = \lambda e^{-\lambda x}$$

and

$$m(t) = 1 - e^{-\lambda t} + \int_0^t m(t-x)\lambda e^{-\lambda x}dx$$

If we make the substitution $t - x = y$, then

$$m(t) = 1 - e^{-\lambda t} + \int_0^t m(y)\lambda e^{-\lambda(t-y)}dy$$

$$m(t) = 1 - e^{-\lambda t} + \lambda e^{-\lambda t}\int_0^t m(y)e^{\lambda y}dy$$

If we multiply both sides by $e^{\lambda t}$, then

$$e^{\lambda t} m(t) = e^{\lambda t} - 1 + \lambda \int_0^t m(y) e^{\lambda y} dy$$

If we differentiate both sides with respect to t, then

$$\lambda e^{\lambda t} m(t) + e^{\lambda t} m'(t) = \lambda e^{\lambda t} + \lambda m(t) e^{\lambda t}$$

and

$$\lambda e^{\lambda t} m(t) + e^{\lambda t} m'(t) = \lambda e^{\lambda t} + \lambda e^{\lambda t} m(t)$$

which reduces to

$$e^{\lambda t} m'(t) = \lambda e^{\lambda t}$$

and

$$m'(t) = \lambda$$

which means

$$m(t) = \lambda t + c$$

Note that

$$m(0) = E[N(0)] = 0$$

$$c = 0$$

Finally,

$$m(t) = E[N(t)] = \lambda t$$

holds for a *homogeneous Poisson process*.

3. Assume that the bus service starts at 7:00 a.m., and the times between the departures of the buses are *i.i.d.* Assume also that you arrive at the bus stop at 11:00 a.m.

(a) Describe the age, excess, and spread of this renewal process.
(b) If we assume that the times between the departures of the buses are *uniformly* distributed over (5, 10) min, how many minutes on average should you wait until departure when you arrive at 11:00 a.m.?

Solution

(a) If we assume 7:00 a.m. as $t = 0$, then the *age* of this renewal process at 11:00 a.m. will be

$$A(4) = 4 - t_{N(4)}$$

the *excess* of this renewal process at 11:00 a.m. will be

$$B(4) = t_{N(4)+1} - 4$$

and the *spread* of this renewal process at 11:00 a.m. will be

$$S(4) = X_{N(4)+1} = t_{N(4)+1} - t_{N(4)} = A(4) + B(4)$$

(b) If we assume that the times between the departures of the buses are *uniformly* distributed over (5,10) min, then when you arrive at 11:00 a.m., you should wait until departure on average

$$\frac{E[X^2]}{2E[X]}$$

minutes by considering the *average excess of a renewal process* from Table 11.2, where

$$E[X] = \frac{5 + 10}{2} = \frac{15}{2}$$

and

$$E[X^2] = Var(X) + (E[X])^2$$
$$= \frac{(10 - 5)^2}{12} + \left(\frac{15}{2}\right)^2 = \frac{175}{3}$$

Finally,

$$\frac{E[X^2]}{2E[X]} = \frac{\left(\frac{175}{3}\right)}{2\left(\frac{15}{2}\right)} = \frac{175}{45} = 3.889 \, \text{min}$$

4. A machine has two components with lifetimes exponentially distributed with rates $\frac{1}{2}$ and $\frac{1}{3}$ per year. Whenever a component breaks down, the machine is stopped, and the broken-down component is repaired. After the repair, the machine becomes like a new machine. If component 1 breaks down, the expected value of the repair time is 1 month, and the expected value of the repair cost is $100; but if component 2 breaks down, the expected value of the repair time is 2 months, and the expected value of the repair cost is $150. This process continues infinitely with *independent* and *identical* distributions.

(a) What is the long-run average rate of the machine renewal?
(b) What is the long-run average cost?
(c) What is the limiting probability that the machine is *on*?

Solution
Let

Z_n: Lifetime of the machine in the nth renewal
Y_n: Repair time of the machine in the nth renewal
X_{1n}: Lifetime of component 1 in the nth renewal
X_{2n}: Lifetime of component 2 in the nth renewal

(a) The long-run average rate of the machine renewal will be

$$\frac{N(t)}{t} \rightarrow \frac{1}{E[Z_n] + E[Y_n]} \quad \text{as } t \rightarrow \infty$$

from *limit theorem*, where

$N(t)$: Number of the machine renewals in $(0, t]$

Thus, we need to compute $E[Z_n]$ and $E[Y_n]$ for $n \geq 1$. Note that

$$Z_n = \min(X_{1n}, X_{2n}), n \geq 1$$

Since

$$X_{1n} \sim \exp\left(\frac{1}{2}\right) \quad n \geq 1$$

$$X_{2n} \sim \exp\left(\frac{1}{3}\right) \quad n \geq 1$$

it follows that

$$Z_n \sim \exp\left(\frac{1}{2} + \frac{1}{3}\right) \text{ and } Z_n \sim \exp\left(\frac{5}{6}\right) \quad n \geq 1$$

Finally,

$$E[Z_n] = \frac{1}{\left(\frac{5}{6}\right)} = 1.2 \text{ years} \quad n \geq 1$$

Note also that $E[Y_n], n \geq 1$
can be computed by conditioning on the first component to fail such that

$$E[Y_n] = E(Y_n|X_{1n} < X_{2n})P(X_{1n} < X_{2n}) + E(Y_n|X_{2n} < X_{1n})P(X_{2n} < X_{1n})$$

$$= \frac{1}{12}\left(\frac{\frac{1}{2}}{\frac{1}{2} + \frac{1}{3}}\right) + \frac{2}{12}\left(\frac{\frac{1}{3}}{\frac{1}{2} + \frac{1}{3}}\right) = \frac{1}{20} + \frac{1}{15} = \frac{7}{60} = 0.117 \text{ year}$$

As a result, as $t \to \infty$,

$$\frac{N(t)}{t} \to \frac{1}{E[Z_n] + E[Y_n]} = \frac{1}{1.2 + 0.117} = \frac{1}{1.317} = 0.7593 \text{ per year}$$

(b) The *long-run average cost* will be

$$\frac{E[\text{cost of a renewal cycle}]}{E[\text{length of a renewal cycle}]}$$

by *renewal reward process*. Note that

$$E[\text{cost of a renewal cycle}]$$
$$= E(\text{cost of a renewal cycle} \mid X_{1n} < X_{2n})P(X_{1n} < X_{2n})$$
$$+ E(\text{cost of a renewal cycle} \mid X_{2n} < X_{1n})P(X_{2n} < X_{1n})$$
$$= 100\left(\frac{\frac{1}{2}}{\frac{1}{2} + \frac{1}{3}}\right) + 150\left(\frac{\frac{1}{3}}{\frac{1}{2} + \frac{1}{3}}\right) = 60 + 60 = \$120$$

Recall that

$$E[\text{length of a renewal cycle}] = E[Z_n] + E[Y_n] = 1.317 \text{ years}$$

from part (a).

Finally, the long-run average cost will be

$$\frac{E[\text{cost of a renewal cycle}]}{E[\text{length of a renewal cycle}]} = \frac{120}{1.317} \cong \$91.116 \text{ per year}$$

(c) The limiting probability that the machine is *on* will be

$$\lim_{t \to \infty} P_{on}(t) = \frac{E[Z_n]}{E[Z_n] + E[Y_n]} = \frac{1.2}{1.2 + 0.117} = 0.9112$$

since this process is an *alternating renewal process*.

Chapter 12
An Introduction to Markov Chains

Abstract This chapter was a compact introduction to both discrete-time and continuous-time Markov chains. The important concepts including the definition of discrete-time and continuous-time Markov chains, Chapman-Kolmogorov equations, reachability, communication, communication classes, recurrent and transient states, period of a state for a discrete-time Markov chain, the limiting probability of a state for a discrete-time and continuous-time Markov chain, and ergodic Markov chain were explained. Several examples and problems have been solved for the discrete-time Markov chains, and where relevant, state transition diagrams and tables have been used to facilitate the comprehension of the solutions.

This chapter will introduce a very important type of stochastic process which is based on conditional probability, i.e. Markov chains. After providing the basic definitions for the discrete-time and continuous-time Markov chains, the important basic properties such as Chapman-Kolmogorov equations, communication of states, recurrent and transient states, period of a state, limiting probability of a state, and ergodic Markov chain will be given. The learnings from this chapter will be background for the Chaps. 13, 14 and 15. The examples and problems of this chapter will be limited to the basic properties of the discrete-time Markov chains; since Chap. 13 will introduce some special Markov chains with some examples and problems, Chap. 14 will be devoted to the continuous-time Markov chains with several examples and problems, and Chap. 15 will be devoted to the queueing models as an extension of the continuous-time Markov chains with illustrative examples and problems.

12.1 Basic Concepts

Example 1 The possible weather conditions in a region for a specific season are assumed to be *sunny*, *rainy*, and *cloudy*. Assume that each day's weather condition depends *only* on the weather condition of the *previous day*.

© Springer Nature Switzerland AG 2019
E. Bas, *Basics of Probability and Stochastic Processes*,
https://doi.org/10.1007/978-3-030-32323-3_12

Definition 1 (*Markovian property*) If a stochastic process has the property that "*the future is independent of the past given the present state*", then it has the *Markovian* property, i.e., the past states are irrelevant for this stochastic process.

Definition 2 (*Discrete-time vs. continuous-time vs. embedded discrete-time Markov chain*) A stochastic process

$$\{X_n; n \geq 0\}$$

is called as a *Discrete-Time Markov Chain* (*DTMC*) if it has the *Markovian property*

$$P(X_{n+1} = j | X_0 = i_0, X_1 = i_1, X_2 = i_2, \ldots, X_n = i) = P(X_{n+1} = j | X_n = i) = P_{ij}$$

where n represents the *present step*, and $n + 1$ represents the *next (future) step*. P_{ij} can be interpreted as "the probability that Markov chain will be in state j *one step later* given that it is in state i now". P_{ij} can also be called as *one-step transition probability* for transition from state i to state j.

A stochastic process

$$\{X(t); t \geq 0\}$$

is called as a *Continuous-Time Markov Chain* (*CTMC*) if it has the *Markovian property*

$$P(X(t + s) = j | X(u), 0 \leq u < s, X(s) = i) = P(X(t + s) = j | X(s) = i) = P_{ij}(t)$$

where $P_{ij}(t)$ can be interpreted as "the probability that Markov chain will be in state j *t time units later* given that it is in state i now". $P_{ij}(t)$ can also be called as *transition probability function* for transition from state i to state j.

A discrete-time Markov chain associated with a *CTMC* is called as an *embedded discrete-time Markov chain* if we consider *only* the *states* of the successive transitions.

Remark 1 In a *CTMC*, the time spent in a state is a *random variable*, whereas the time spent in a state is assumed to be certain in a *DTMC*.

Definition 3 (*Stationary property, time-homogeneity*) In a *DTMC*, if $P(X_{n+1} = j | X_n = i) = P_{ij}$ holds for all $n \geq 0$ values, and for all states i, j, then *DTMC* has the *stationary property* or *time-homogeneity*. Likewise, in a *CTMC*, if $P(X(t + s) = j | X(s) = i) = P_{ij}(t)$ holds for all $s \geq 0, t > 0$ values, and for all states i, j, then *CTMC* has the *stationary property* or *time-homogeneity*.

Remark 2 Analogous to the *stationary increments* in a Poisson process, where the *length* of the interval is relevant for distribution, the *number of steps* or the *number of time units* is relevant for distribution in a Markov chain in case of *stationary property*.

Definition 4 (*State space for Markov chains*) The set of all possible states of a random variable of a Markov chain is called as a *state space*, and denoted by S. The set of nonnegative integer numbers $S = \{0, 1, 2, 3, \ldots\}$ is generally assumed to be the state space for a Markov chain.

Remark 3 Although the states of a Markov chain can be *verbal*, they can be converted to the *nonnegative integer numbers*.

Example 1 (revisited 1) Recall that the possible weather conditions in a region for a specific season are assumed to be *sunny*, *rainy*, and *cloudy*. If the weather *on any day* depends *only* on the weather of the *previous day*, then the weather conditions can be represented as a *DTMC* such that

$$\{X_n; n \geq 0\}$$
$$X_n : \text{Weather condition on the } n\text{th day, and}$$
$$S = \{0, 1, 2\}$$
$$0 : \text{Sunny}$$
$$1 : \text{Rainy}$$
$$2 : \text{Cloudy}$$

As an example, "the probability that it will be *rainy tomorrow* given that it is *cloudy today* is 0.6" can be denoted as

$$P_{21} = 0.6$$

If this probability does not change for any two successive days, then this Markov chain has the *stationary property*. As an example, in case of *stationary property*, the probability that it will be *rainy* on Wednesday given that it is *cloudy* on Tuesday or the probability that it will be *rainy* on Sunday given that it is *cloudy* on Saturday will be 0.6.

12.2 Transition Probability Matrix, Matrix of Transition Probability Functions, Chapman-Kolmogorov Equations

Definition 5 (*One-step transition probability matrix vs. matrix of transition probability functions*) Table 12.1 provides the definitions of the one-step transition probability matrix, and the matrix of transition probability functions in comparison to each other.

Table 12.1 One-step transition probability matrix versus matrix of transition probability functions

Discrete-time Markov chain	Continuous-time Markov chain
For a $DTMC$ with state space $S = \{0, 1, 2, \ldots, K\}$, $$P = \begin{bmatrix} P_{00} & \cdots & P_{0K} \\ \vdots & P_{ij} & \vdots \\ P_{K0} & \cdots & P_{KK} \end{bmatrix}$$ is the one-step transition probability matrix, where P_{ij} : One-step transition probability for transition from state i to state j, and $\sum_{j \in S} P_{ij} = 1 \quad \forall i \in S$	For a $CTMC$ with state space $S = \{0, 1, 2, \ldots, K\}$, $$P(t) = \begin{bmatrix} P_{00}(t) & \cdots & P_{0K}(t) \\ \vdots & P_{ij}(t) & \vdots \\ P_{K0}(t) & \cdots & P_{KK}(t) \end{bmatrix}$$ is the matrix of transition probability functions, where $P_{ij}(t)$: Transition probability function for transition from state i to state j, and $\sum_{j \in S} P_{ij}(t) = 1 \quad \forall i \in S$

Example 1 (revisited 2) We assume the *one-step* transition probability matrix as

$$P = \begin{bmatrix} 0.3 & 0.4 & 0.3 \\ 0.5 & 0.2 & 0.3 \\ 0.2 & 0.6 & 0.2 \end{bmatrix}$$

As an example, the probability that "it will be *sunny* tomorrow, given that it is *rainy* today" will be $P_{10} = 0.5$. Note that the sum of each row is 1.

Definition 6 (*Chapman-Kolmogorov equations, n-step transition probability matrix, matrix of transition probability functions*) Table 12.2 provides the definitions of Chapman-Kolmogorov equations for the discrete-time and continuous-time Markov chains, the *n*-step transition probability matrix, and the matrix of transition probability functions.

Example 1 (revisited 3) Consider the *two-step* transition probability matrix of the problem as

$$P^{(2)} = P^2 = \begin{bmatrix} 0.3 & 0.4 & 0.3 \\ 0.5 & 0.2 & 0.3 \\ 0.2 & 0.6 & 0.2 \end{bmatrix} \begin{bmatrix} 0.3 & 0.4 & 0.3 \\ 0.5 & 0.2 & 0.3 \\ 0.2 & 0.6 & 0.2 \end{bmatrix} = \begin{bmatrix} 0.35 & 0.38 & 0.27 \\ 0.31 & 0.42 & 0.27 \\ 0.40 & 0.32 & 0.28 \end{bmatrix}$$

As an example, $P_{10}^2 = 0.31$ is the probability that it will be sunny *two days later* given that it is rainy *today*. Note again that the sum of each row is 1.

Table 12.2 Chapman-Kolmogorov equations, n-step transition probability matrix, and matrix of transition probability functions

Discrete-time Markov chain	Continuous-time Markov chain
For a $DTMC$ with state space $S = \{0, 1, 2, \ldots, K\}$, $$P_{ij}^{n+m} = \sum_{k=0}^{K} P_{ik}^{n} P_{kj}^{m} \quad \text{for } n, m \geq 0$$ for all $i, j \in S$ equations are called as *Chapman-Kolmogorov equations*, and	For a $CTMC$ with state space $S = \{0, 1, 2, \ldots, K\}$, $$P_{ij}(t+s) = \sum_{k=0}^{K} P_{ik}(t) P_{kj}(s) \quad \text{for } t, s \geq 0$$ for all $i, j \in S$ equations are called as *Chapman-Kolmogorov equations*, and
$$P^{(n)} = P^n = \begin{bmatrix} P_{00}^{n} & \cdots & P_{0K}^{n} \\ \vdots & P_{ij}^{n} & \vdots \\ P_{K0}^{n} & \cdots & P_{KK}^{n} \end{bmatrix}$$ is the n-step transition probability matrix that follows from Chapman-Kolmogorov equations, where P_{ij}^{n} : n-step transition probability for transition from state i to state j, and $$\sum_{j \in S} P_{ij}^{n} = 1 \quad \forall i \in S$$	$$P(t+s) = P(t)P(s) =$$ $$\begin{bmatrix} P_{00}(t+s) & \cdots & P_{0K}(t+s) \\ \vdots & P_{ij}(t+s) & \vdots \\ P_{K0}(t+s) & \cdots & P_{KK}(t+s) \end{bmatrix}$$ is the matrix of transition probability functions that follows from Chapman-Kolmogorov equations, where $P_{ij}(t+s)$: Transition probability function for transition from state i to state j, and $$\sum_{j \in S} P_{ij}(t+s) = 1 \quad \forall i \in S$$

12.3 Communication Classes, Irreducible Markov Chain, Recurrent Versus Transient States, Period of a State, Ergodic State

Definition 7 (*Communication, communication classes, irreducible Markov chain*) Table 12.3 provides the definitions of communication, communication classes, and irreducible Markov chain.

Remark 4 Note that the states i and j of a $CTMC$ *communicate* with each other if and only if they communicate with each other for the *embedded discrete-time Markov chain*. Note also that a $CTMC$ is *irreducible* if and only if its *embedded discrete-time Markov chain* is *irreducible*.

Example 1 (revisited 4) Recall transition probability matrix for the problem as

$$P = \begin{bmatrix} 0.3 & 0.4 & 0.3 \\ 0.5 & 0.2 & 0.3 \\ 0.2 & 0.6 & 0.2 \end{bmatrix}$$

Table 12.3 Communication, communication classes, irreducible Markov chain

Discrete-time Markov chain	Continuous-time Markov chain
State j is said to be *reachable* from state i if $P_{ij}^n > 0 \quad n \geq 0$ and it is denoted by $i \rightarrow j$ If $i \rightarrow j$ and $j \rightarrow i$, then state i and state j are said to *communicate* with each other, and it is denoted by $i \longleftrightarrow j$ Each state space of a $DTMC$ is the union of so-called *communication classes* such that $S = C_1 \cup C_2 \cup \ldots \cup C_r$ $C_1 \cap C_2 \cap \ldots \cap C_r = \emptyset$ Note that each state in a communication class communicates with each other, but not with the other states in other communication classes. If $S = C$, then $DTMC$ is said to be *irreducible*, otherwise *reducible*.	State j is said to be *reachable* from state i if $P_{ij}(t) > 0 \quad t \geq 0$ and it is denoted by $i \rightarrow j$ If $i \rightarrow j$ and $j \rightarrow i$, then state i and state j are said to *communicate* with each other, and it is denoted by $i \longleftrightarrow j$ Each state space of a $CTMC$ is the union of so-called *communication classes* such that $S = C_1 \cup C_2 \cup \ldots \cup C_r$ $C_1 \cap C_2 \cap \ldots \cap C_r = \emptyset$ Note that each state in a communication class communicates with each other, but not with the other states in other communication classes. If $S = C$, then $CTMC$ is said to be *irreducible*, otherwise *reducible*.

and state space as

$$S = \{0, 1, 2\}$$

Hence, the *state transition diagram* in Fig. 12.1 can be constructed for this
$DTMC$.

The reachability and communication of the states will be as provided in Table 12.4.
Since all states communicate with each other,

$$S = C = \{0, 1, 2\}$$

Fig. 12.1 State transition
diagram for example 1
(revisited 4)

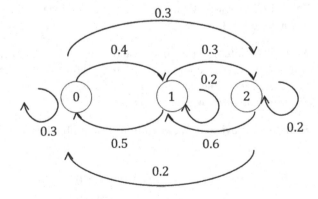

Table 12.4 Reachability and communication of the states for example 1 (revisited 4)	$0 \longleftrightarrow 1$	$1 \longleftrightarrow 2$
	$0 \longleftrightarrow 2$	

and this Markov chain is *irreducible*.

Definition 8 (*Absorbing state for a discrete-time Markov chain*) State i of a discrete-time Markov chain is an absorbing state if

$$P_{ii} = 1$$

Remark 5 An absorbing state for a discrete-time Markov chain *always* forms a *communication class only* with itself. A state of a $CTMC$ is an absorbing state when no transition out of this state is possible.

Example 1 (revisited 5) If the transition probability matrix is modified as

$$P = \begin{bmatrix} 0.3 & 0.4 & 0.3 \\ 0 & 1 & 0 \\ 0.2 & 0.6 & 0.2 \end{bmatrix}$$

then $P_{11} = 1$ which means *state 1*, rainy weather, is an *absorbing* state. In this case, the *state transition diagram* will be as in Fig. 12.2.

The reachability and communication of the states will be as in Table 12.5. Since

Fig. 12.2 State transition diagram for example 1 (revisited 5)

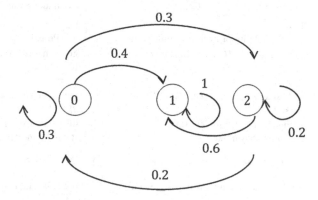

Table 12.5 Reachability and communication of the states for example 1 (revisited 5)	$0 \rightarrow 1$	$1 \leftarrow 2$
	$0 \longleftrightarrow 2$	

$$C_1 = \{1\}$$
$$C_2 = \{0, 2\}$$
$$S = C_1 \cup C_2$$

this Markov chain is *reducible*.

Remark 6 Note that although state 1 is reachable from state 0 and from state 2, state 0 and state 2 are *not* reachable from state 1, which means states 0,1 and states 1,2 do *not* communicate with each other. Note also that *absorbing state* 1 forms a communication class *only* with *itself*.

Definition 9 (*Recurrent vs. transient states*) Table 12.6 provides the definitions of the recurrent and transient states for the discrete-time and continuous-time Markov chains.

Remark 7 A Markov chain visits a *recurrent state infinitely* again and again, while it visits a *transient state* only for a *finite number* of times. An *absorbing* state is always a *recurrent* state for a *DTMC*. The recurrence and transience are the *class properties*, that is, when a state in a communication class is recurrent (transient), all other states in this communication class are also recurrent (transient). Note that a state of a *CTMC* is recurrent (transient) if and only if it is recurrent (transient) for the *embedded discrete-time Markov chain*.

Example 1 (revisited 6) If the transition probability matrix is assumed to be

Table 12.6 Recurrent versus transient states

Discrete-time Markov Chain	Continuous-time Markov Chain
Let	Let
f_i : Probability of ever returning to state i given that $DTMC$ starts in state i	f_i : Probability of ever returning to state i given that $CTMC$ starts in state i
If	If
$f_i = 1$	$f_i = 1$
then state i is a *recurrent* state. If	then state i is a *recurrent* state. If
$f_i < 1$	$f_i < 1$
then state i is a *transient* state.	then state i is a *transient* state.
It can be shown that	It can be shown that
$\displaystyle\sum_{n=1}^{\infty} P_{ii}^n = \infty$	$\displaystyle\sum_{n=1}^{\infty} P_{ii}^n = \infty$
holds if state i is a *recurrent* state, and	holds if state i is a *recurrent* state, and
$\displaystyle\sum_{n=1}^{\infty} P_{ii}^n < \infty$	$\displaystyle\sum_{n=1}^{\infty} P_{ii}^n < \infty$
holds if state i is a *transient* state.	holds if state i is a *transient* state, where n is the number of transitions for the *embedded discrete-time Markov chain* of a *CTMC*.

$$P = \begin{bmatrix} 0.3 & 0.4 & 0.3 \\ 0 & 1 & 0 \\ 0.2 & 0.6 & 0.2 \end{bmatrix}$$

then

$$C_1 = \{1\}$$
$$C_2 = \{0, 2\}$$

will be the communication classes as given in Example 1 (revisited 5), where state 1 is an *absorbing* and a *recurrent* state, while the states 0 and 2 are the *transient* states. Note that once Markov chain jumps from state 0 or from state 2 to state 1, it cannot return to state 0 or state 2, thus, the states 0 and 2 are the transient states.

Definition 10 (*Positive recurrent vs. null recurrent state for a discrete-time Markov chain*) Consider a $DTMC$

$$\{X_n; n \geq 0\}$$

and let

$$\tau_{ii} = min\{n \geq 1 : X_n = i | X_0 = i\}$$

which means

τ_{ii}: Minimum number of steps to return to state i given that $DTMC$ starts in state i

Accordingly, the positive recurrent state and null recurrent state can be defined as provided in Table 12.7.

Remark 8 A recurrent state of a $CTMC$ is positive (null) recurrent if and only if it is positive (null) recurrent for the *embedded discrete-time Markov chain*.

Definition 11 (*Period of a state, aperiodic state for a discrete-time Markov chain*) State i of a discrete-time Markov chain is said to have period $d(i)$ such that

$$d(i) = g.c.d.\{n > 0 | P_{ii}^n > 0\}$$

where *g.c.d.* stands for *greatest common divisor*. State i is called as *aperiodic* if $d(i) = 1$.

Table 12.7 Positive recurrent state versus null recurrent state

Positive recurrent state	Null recurrent state
A recurrent state i is *positive recurrent* if $E[\tau_{ii}] < \infty$	A recurrent state i is *null recurrent* if $E[\tau_{ii}] = \infty$

Fig. 12.3 State transition
diagram for example 2

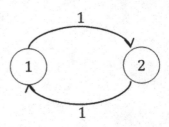

Remark 9 A period of a state is also a *class property*. If state i and state j are in the same communication class, then $d(i) = d(j)$. Since there are no time steps for a continuous-time Markov chain, periodicity is *not defined* for a continuous-time Markov chain.

Example 2 Let $\{X_n; n \geq 0\}$ be a Markov chain with $S = \{1, 2\}$ and with transition probability matrix

$$P = \begin{bmatrix} 0 & 1 \\ 1 & 0 \end{bmatrix}$$

We consider the *state transition diagram* in Fig. 12.3 for this Markov chain. Accordingly,

$$d(1) = d(2) = 2$$

since

$$P_{11}^2 > 0, \, P_{11}^4 > 0, \, P_{11}^6 > 0, \, P_{11}^8 > 0, \ldots$$

$$P_{22}^2 > 0, \, P_{22}^4 > 0, \, P_{22}^6 > 0, \, P_{22}^8 > 0, \ldots$$

which means

$$d(1) = d(2) = g.c.d.\{2, 4, 6, 8, \ldots\} = 2$$

If the transition probability matrix were modified as

$$P = \begin{bmatrix} 1/3 & 2/3 \\ 1/2 & 1/2 \end{bmatrix}$$

then

$$d(1) = d(2) = 1$$

would hold since

$$P_{11} > 0, P_{11}^2 > 0, P_{11}^3 > 0, P_{11}^4 > 0, \ldots$$

$$P_{22} > 0, P_{22}^2 > 0, P_{22}^3 > 0, P_{22}^4 > 0, \ldots$$

which means $d(1) = d(2) = g.c.d.\{1, 2, 3, 4, \ldots\} = 1$.

Definition 12 (*Ergodic state for a discrete-time Markov chain*) If a state of a $DTMC$ is *positive recurrent* and *aperiodic*, then this state is called as an *ergodic state*.

12.4 Limiting Probability of a State of a Markov Chain

Definition 13 (*Limiting probability of a state*) The limiting probability of a state for a $DTMC$ and $CTMC$ can be defined as in Table 12.8.

Table 12.8 Limiting probability of a state

Discrete-time Markov Chain	Continuous-time Markov Chain
Consider an *irreducible DTMC* with state space $S = \{0, 1, 2, \ldots, K\}$ where all states are *positive recurrent*. If $\pi_i = \lim\limits_{n \to \infty} P_{ji}^n \quad i \in S$ exists and independent of the initial state j, then π_i is called as the *limiting probability* or the *steady-state probability of state i* or the *long-run proportion of time that DTMC is in state i*. Note that $\sum\limits_{i=0}^{K} \pi_i = 1$ and π_i is the unique solution of $\pi_i = \sum\limits_{j=0}^{K} \pi_j P_{ji} \quad i \in S$ which are called as the *balance equations*	Consider an *irreducible CTMC* with state space $S = \{0, 1, 2, \ldots, K\}$ where all states are *positive recurrent*. If $P_i = \lim\limits_{t \to \infty} P_{ji}(t) \quad i \in S$ exists and independent of the initial state j, then P_i is called as the *limiting probability* or the *steady-state probability of state i* or the *long-run proportion of time that CTMC is in state i*. Note that $\sum\limits_{i=0}^{K} P_i = 1$ and P_i is the unique solution of $v_i P_i = \sum\limits_{j \neq i, j \in S} q_{ji} P_j \quad i \in S$ which are called as the *balance equations*, where $v_i P_i$: Rate at which *CTMC leaves* state i $\sum\limits_{j \neq i, j \in S} q_{ji} P_j$: Rate at which *CTMC enters* state i v_i: Transition rate for transition from state i q_{ji}: Instantaneous transition rate for transition from state j to state i

Remark 10 For the details of the limiting probability of a state for a *CTMC*, please refer to Chaps. 14 and 15.

Definition 14 (*Ergodic Markov chain, Ergodic theorem*) A discrete-time or continuous-time Markov chain with the properties defined in Table 12.8 is called as an *ergodic Markov chain*.

Remark 11 If all states of a *DTMC* are *positive recurrent* and *aperiodic*, then these states are *ergodic states* as defined in Definition 12, and this *DTMC* is an *ergodic DTMC*.

Example 1 (revisited 7) Recall the transition probability matrix as

$$P = \begin{bmatrix} 0.3 & 0.4 & 0.3 \\ 0.5 & 0.2 & 0.3 \\ 0.2 & 0.6 & 0.2 \end{bmatrix}$$

and the state space as $S = \{0, 1, 2\}$.

Since this *DTMC* is an *irreducible* and *ergodic* Markov chain, then the following *balance equations*

$$\pi_0 = 0.3\pi_0 + 0.5\pi_1 + 0.2\pi_2 \tag{1}$$

$$\pi_1 = 0.4\pi_0 + 0.2\pi_1 + 0.6\pi_2 \tag{2}$$

$$\pi_2 = 0.3\pi_0 + 0.3\pi_1 + 0.2\pi_2 \tag{3}$$

and the following equation

$$\pi_0 + \pi_1 + \pi_2 = 1 \tag{4}$$

should hold.

We can choose any two equations out of the balance equations(1–3) since they are *linearly dependent*. As an example, we choose Eqs. (1), (2), and also by using Eq. (4), we can find

$$\pi_0 = \frac{23}{66} \quad \pi_1 = \frac{25}{66} \quad \pi_2 = \frac{18}{66}$$

Problems

1. Consider the maze in Fig. 12.4 through which a rat can move. The rat has no memory. At each time, it can only move to the neighboring cells with equal probability, and it is possible to go to freedom only from cell 4 with the same probability of moving to the neighboring cells. If the rat goes to freedom, it stays in this state forever.

Fig. 12.4 Rat in the maze
problem

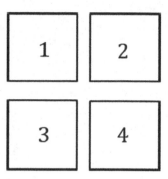

(a) Describe the movements of the rat as a *DTMC*.
(b) Build the transition probability matrix and the state transition diagram for this *DTMC*.
(c) Identify the communication classes, the recurrent and the transient states.
(d) Find the limiting probability of each state.

Solution

(a) $\{X_n; n \geq 0\}$ is a *DTMC*

since the current cell of the rat depends on the previous cell, where

X_n : The cell of the rat at the nth movement

$$S = \{0, 1, 2, 3, 4\}$$

if we consider state 0 as freedom.

(b) The following transition probability matrix, and the state transition diagram given in Fig. 12.5 can be built for this *DTMC*.

$$P = \begin{bmatrix} 1 & 0 & 0 & 0 & 0 \\ 0 & 0 & 1/2 & 1/2 & 0 \\ 0 & 1/2 & 0 & 0 & 1/2 \\ 0 & 1/2 & 0 & 0 & 1/2 \\ 1/3 & 0 & 1/3 & 1/3 & 0 \end{bmatrix}$$

(c) The reachability and communication of the states will be as in Table 12.9.

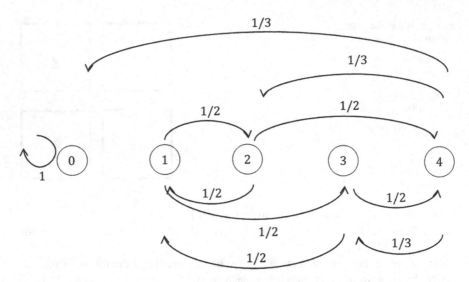

Fig. 12.5 State transition diagram for the rat in the maze problem

Table 12.9 Reachability and	$0 \leftarrow 1$	$1 \longleftrightarrow 2$	$2 \longleftrightarrow 3$	$3 \longleftrightarrow 4$
communication of the states for the rat in the maze problem	$0 \leftarrow 2$	$1 \longleftrightarrow 3$	$2 \longleftrightarrow 4$	
	$0 \leftarrow 3$	$1 \longleftrightarrow 4$		
	$0 \leftarrow 4$			

Since

$$C_1 = \{0\}$$
$$C_2 = \{1, 2, 3, 4\}$$
$$S = C_1 \cup C_2$$

this Markov chain is a *reducible* Markov chain. State 0 is an *absorbing* and a *recurrent* state, whereas the states 1, 2, 3, 4 are the *transient* states.

(d) Since this Markov chain is a *reducible* Markov chain, the limiting probabilities of the states cannot be computed.

WARNING Reachability from one state to another state does not have to be in one step. As an example, state 3 can be reached from state 2 via state 1 or state 4, and state 2 can be reached from state 3 via state 1 or state 4. Although the states 1, 2, 3, 4 communicate with each other, thus form a communication class, once the Markov chain jumps from state 4 to state 0, it stays in that state forever. Thus, the states 1, 2, 3, 4 are visited for a finite number of times, and as a result they are the *transient* states.

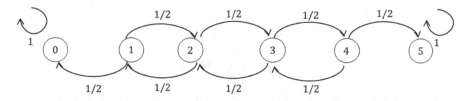

Fig. 12.6 State transition diagram for problem 2

2. Consider a system with state space $S = \{0, 1, 2, 3, 4, 5\}$. Whenever the system is in state i for $0 < i < 5$, at the next step, it will enter into state $i + 1$ or $i - 1$ with equal probability. When the system enters into state 0 or state 5, it will remain there forever.

   (a) Build the state transition diagram and the transition probability matrix for this *DTMC*.
   (b) Identify the communication classes, the recurrent and the transient states.
   (c) Find the limiting probability of each state.

Solution

(a) The system can be described with the state transition diagram given in Fig. 12.6. The system can also be described with the following transition probability matrix:

$$P = \begin{bmatrix} 1 & 0 & 0 & 0 & 0 & 0 \\ 1/2 & 0 & 1/2 & 0 & 0 & 0 \\ 0 & 1/2 & 0 & 1/2 & 0 & 0 \\ 0 & 0 & 1/2 & 0 & 1/2 & 0 \\ 0 & 0 & 0 & 1/2 & 0 & 1/2 \\ 0 & 0 & 0 & 0 & 0 & 1 \end{bmatrix}$$

(b) The reachability and communication of the states will be as in Table 12.10.

   Since

Table 12.10 Reachability and communication of the states for problem 2

$0 \leftarrow 1$	$1 \longleftrightarrow 2$	$2 \longleftrightarrow 3$	$3 \longleftrightarrow 4$	$4 \rightarrow 5$
$0 \leftarrow 2$	$1 \longleftrightarrow 3$	$2 \longleftrightarrow 4$	$3 \rightarrow 5$	
$0 \leftarrow 3$	$1 \longleftrightarrow 4$	$2 \rightarrow 5$		
$0 \leftarrow 4$	$1 \rightarrow 5$			
$0 \quad 5$				

$$C_1 = \{0\}$$
$$C_2 = \{1, 2, 3, 4\}$$
$$C_3 = \{5\}$$
$$S = C_1 \cup C_2 \cup C_3$$

this Markov chain is *reducible*. The states 0 and 5 are the *absorbing* and the *recurrent* states, whereas the states 1, 2, 3, 4 are the *transient* states.

(c) Since this Markov chain is *reducible*, the limiting probabilities of the states cannot be computed.

3. Consider the Markov chain in Example 1 (revisited 2) with state space $S = \{0, 1, 2\}$ and with following transition probability matrix

$$P = \begin{bmatrix} 0.3 \ 0.4 \ 0.3 \\ 0.5 \ 0.2 \ 0.3 \\ 0.2 \ 0.6 \ 0.2 \end{bmatrix}$$

If $P(X_0 = 1) = P(X_0 = 2) = \frac{1}{5}$, then find $E[X_2]$. Find also the expected value of the states in the long-run.

Solution

Note that

$$E[X_2] = (0)P(X_2 = 0) + (1)P(X_2 = 1) + (2)P(X_2 = 2)$$
$$= P(X_2 = 1) + 2P(X_2 = 2)$$

where

$$P(X_2 = 1) = P(X_2 = 1|X_0 = 0)P(X_0 = 0) + P(X_2 = 1|X_0 = 1)P(X_0 = 1)$$
$$+ P(X_2 = 1|X_0 = 2)P(X_0 = 2)$$
$$= P_{01}^2 P(X_0 = 0) + P_{11}^2 P(X_0 = 1) + P_{21}^2 P(X_0 = 2)$$

Analogously,

$$P(X_2 = 2) = P_{02}^2 P(X_0 = 0) + P_{12}^2 P(X_0 = 1) + P_{22}^2 P(X_0 = 2)$$

It is clear that

$$P(X_0 = 0) = 1 - [P(X_0 = 1) + P(X_0 = 2)] = 1 - \left[\frac{1}{5} + \frac{1}{5}\right] = \frac{3}{5}$$

Recall the *two-step* transition probability matrix for this $DTMC$ from Example 1 (revisited 3) as

$$P^{(2)} = P^2 = \begin{bmatrix} 0.35 & 0.38 & 0.27 \\ 0.31 & 0.42 & 0.27 \\ 0.40 & 0.32 & 0.28 \end{bmatrix}$$

Thus,

$$P(X_2 = 1) = P_{01}^2 P(X_0 = 0) + P_{11}^2 P(X_0 = 1) + P_{21}^2 P(X_0 = 2)$$
$$= 0.38\left(\frac{3}{5}\right) + 0.42\left(\frac{1}{5}\right) + 0.32\left(\frac{1}{5}\right) = 0.376$$

$$P(X_2 = 2) = P_{02}^2 P(X_0 = 0) + P_{12}^2 P(X_0 = 1) + P_{22}^2 P(X_0 = 2)$$
$$= 0.27\left(\frac{3}{5}\right) + 0.27\left(\frac{1}{5}\right) + 0.28\left(\frac{1}{5}\right) = 0.272$$

Finally,

$$E[X_2] = P(X_2 = 1) + 2P(X_2 = 2)$$
$$= 0.376 + 2(0.272) = 0.92$$

The expected value of the states in the long-run will be

$$E[X] = 0\pi_0 + 1\pi_1 + 2\pi_2$$

Since the limiting probabilities for this $DTMC$ are

$$\pi_0 = \frac{23}{66}, \pi_1 = \frac{25}{66}, \pi_2 = \frac{18}{66}$$

from Example 1 (revisited 7),

$$E[X] = 0\left(\frac{23}{66}\right) + 1\left(\frac{25}{66}\right) + 2\left(\frac{18}{66}\right) = \frac{61}{66} = 0.9242$$

4. Please consider the following transition probability matrices for the Markov chains with sample space $S = \{0, 1, 2\}$:

(a) $P = \begin{bmatrix} 1/2 & 0 & 1/2 \\ 1/2 & 1/2 & 0 \\ 0 & 0 & 1 \end{bmatrix}$

(b) $P = \begin{bmatrix} 1/3 & 2/3 & 0 \\ 1/2 & 1/2 & 0 \\ 0 & 1/4 & 3/4 \end{bmatrix}$

Fig. 12.7 State transition
diagram for problem 4(a)

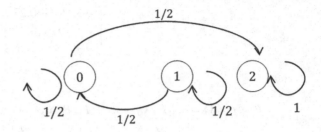

Table 12.11 Reachability
and communication of the
states for problem 4(a)

$0 \leftarrow 1$	$1 \rightarrow 2$
$0 \rightarrow 2$	

(c) $P = \begin{bmatrix} 0 & 0 & 1 \\ 1 & 0 & 0 \\ 0 & 1/2 & 1/2 \end{bmatrix}$

Build the state transition diagram for each Markov chain. Identify the recurrent
and the transient states, and determine whether each Markov chain is *irreducible* or
reducible. Find also the period of the states of each Markov chain, and determine
whether these Markov chains are *ergodic* or not.

Solution

(a) The state transition diagram for this Markov chain will be as in Fig. 12.7.

The reachability and communication of the states will be as in Table 12.11.
Since

$$C_1 = \{0\}$$
$$C_2 = \{1\}$$
$$C_3 = \{2\}$$
$$S = C_1 \cup C_2 \cup C_3$$

this Markov chain is *reducible*. The states 0 and 1 are the *transient* states, whereas
state 2 is an *absorbing* and a *recurrent* state.
Since $P_{00} > 0$, $P_{00}^2 > 0$, $P_{00}^3 > 0$, $P_{00}^4 > 0, \ldots$

$$d(0) = 1$$

Since $P_{11} > 0$, $P_{11}^2 > 0$, $P_{11}^3 > 0$, $P_{11}^4 > 0, \ldots$

$$d(1) = 1$$

Since $P_{22} > 0$, $P_{22}^2 > 0$, $P_{22}^3 > 0$, $P_{22}^4 > 0, \ldots$

Fig. 12.8 State transition
diagram for problem 4(b)

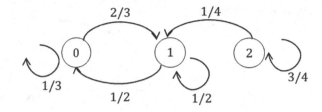

Table 12.12 Reachability
and communication of the
states for problem 4(b)

$0 \longleftrightarrow 1$	$1 \leftarrow 2$
$0 \leftarrow 2$	

$$d(2) = 1$$

Hence, all states are *aperiodic* states. However, since this Markov chain is a
reducible Markov chain with one recurrent state and two transient states, it is *not an
ergodic* Markov chain.

(b) The state transition diagram for this Markov chain will be as in Fig. 12.8.

The reachability and communication of the states will be as in Table 12.12.
Since

$$C_1 = \{0, 1\}$$
$$C_2 = \{2\}$$
$$S = C_1 \cup C_2$$

this Markov chain is *reducible*. The states 0 and 1 are the *recurrent* states, whereas
state 2 is a *transient* state.

Since $P_{00} > 0,\ P_{00}^2 > 0,\ P_{00}^3 > 0,\ P_{00}^4 > 0, \ldots$

$$d(0) = 1$$

Since $P_{11} > 0,\ P_{11}^2 > 0,\ P_{11}^3 > 0,\ P_{11}^4 > 0, \ldots$

$$d(1) = 1$$

Since $P_{22} > 0,\ P_{22}^2 > 0,\ P_{22}^3 > 0,\ P_{22}^4 > 0, \ldots$

$$d(2) = 1$$

Hence, all states are *aperiodic* states. However, since this Markov chain is a
reducible Markov chain with two recurrent states and one transient state, it is *not an
ergodic* Markov chain.

(c) The state transition diagram for this Markov chain will be as in Fig. 12.9.

Fig. 12.9 State transition
diagram for problem 4(c)

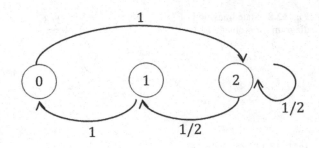

Table 12.13 Reachability
and communication of the
states for problem 4(c)

$0 \longleftrightarrow 1$	$1 \longleftrightarrow 2$
$0 \longleftrightarrow 2$	

The reachability and communication of the states will be as in Table 12.13.
Since

$$S = C = \{0, 1, 2\}$$

this Markov chain is an *irreducible* Markov chain with *all* states *recurrent*.
Since $P_{00}^3 > 0$, $P_{00}^4 > 0$, $P_{00}^5 > 0, \ldots$

$$d(0) = 1$$

Since $P_{11}^3 > 0$, $P_{11}^4 > 0$, $P_{11}^5 > 0, \ldots$

$$d(1) = 1$$

Since $P_{22} > 0$, $P_{22}^2 > 0$, $P_{22}^3 > 0$, $P_{22}^4 > 0, \ldots$

$$d(2) = 1$$

Hence, all states are *aperiodic* states. Since this Markov chain is an *irreducible*
Markov chain with all states *aperiodic* and *positive recurrent*, it is *an ergodic* Markov
chain.

Chapter 13
Special Discrete-Time Markov Chains

Abstract This chapter has been devoted to some selected special discrete-time Markov chains with wide applications including random walk, simple random walk, simple symmetric random walk, Gambler's ruin problem as a special simple random walk, branching process, hidden Markov chains, time-reversible discrete-time Markov chains, and Markov decision processes. Some illustrative examples and problems have been provided for each special discrete-time Markov chain.

After providing the basic definitions and the basic properties for the discrete-time Markov chains in Chap. 12, this chapter will introduce some special and very important $DTMCs$ with representative examples and problems, i.e. random walk, simple random walk, simple symmetric random walk, Gambler's ruin problem as a special simple random walk, branching process including the probability of the eventual population extinction, hidden Markov chains, time-reversible discrete-time Markov chains, and Markov decision processes. Conditional probability again plays an important role in this chapter, and some basic formulae such as multiplication rule should be recalled from Chap. 3. The definitions related to the random walk will be background for Brownian motion in Chap. 16, which is a continuous-time extension of a simple symmetric random walk.

13.1 Random Walk

Definition 1 (*Random walk*) Let $\{\Delta_i; i \geq 1\}$ be *any i.i.d.* sequence. If we define

$$X_n = X_0 + \Delta_1 + \Delta_2 + \cdots + \Delta_i + \cdots + \Delta_n, \quad \text{where } X_0 = k, \ k \geq 0$$

then

$$X_{n+1} = X_0 + \Delta_1 + \Delta_2 + \cdots + \Delta_i + \cdots + \Delta_n + \Delta_{n+1} = X_n + \Delta_{n+1}$$

© Springer Nature Switzerland AG 2019
E. Bas, *Basics of Probability and Stochastic Processes*,
https://doi.org/10.1007/978-3-030-32323-3_13

Table 13.1 Simple random walk versus simple symmetric random walk

Simple random walk	Simple symmetric random walk
Let	Let
$\{X_n; n \geq 0\}$	$\{X_n; n \geq 0\}$
be a random walk, where	be a random walk, where
$X_n = X_0 + \Delta_1 + \Delta_2 + \cdots + \Delta_i + \cdots + \Delta_n,$	$X_n = X_0 + \Delta_1 + \Delta_2 + \cdots + \Delta_i + \cdots + \Delta_n,$
$X_0 = k, k \geq 0$	$X_0 = k, k \geq 0$
If for all $i \geq 1,$	If for all $i \geq 1,$
$\Delta_i = 1$ with probability p	$\Delta_i = 1$ with probability $1/2$
or	or
$\Delta_i = -1$ with probability $1 - p,$	$\Delta_i = -1$ with probability $1/2,$
then $\{X_n; n \geq 0\}$ is a *simple random walk.*	then $\{X_n; n \geq 0\}$ is a *simple symmetric random walk.*
Note that for all $i \geq 1,$	Note that for all $i \geq 1,$
$E[\Delta_i] = 2p - 1$	$E[\Delta_i] = 0$
$Var(\Delta_i) = 4p(1 - p)$	$Var(\Delta_i) = 1$

will hold. Since X_{n+1} (*future* state) depends on X_n (*present* state) and an *independent* random variable Δ_{n+1} for any $n \geq 0,$

$$\{X_n; n \geq 0\}$$

will be *a special DTMC* called as *random walk*. Note that the state space for a random walk is generally defined as $S = \{\ldots, -2, -1, 0, 1, 2, \ldots\}.$

Definition 2 (*Simple random walk vs. simple symmetric random walk*) The *simple random* walk and *simple symmetric* random walk can be defined in comparison to each other as given in Table 13.1.

Remark 1 A *simple* random walk and a *simple symmetric* random walk can be interpreted as the position of an object which moves with *step size* 1 at each *discrete* time point, as an example, at *every minute*. Note that $+1$ may mean moving *forwards* with step size 1, and -1 may mean moving *backwards* with step size 1.

Definition 3 (*Gambler's ruin problem*) Gambler's ruin problem is a special *simple random walk* $\{X_n; n \geq 0\},$ where

$$X_0 = k, \quad k > 0$$

In a gambler's ruin problem, a gambler starts with $\$k$, plays successively, and at each play, he *wins* $\$1$ with probability p *or loses* $\$1$ with probability $1 - p,$ and

$$X_n = X_0 + \Delta_1 + \Delta_2 + \cdots + \Delta_n$$

will be the *total fortune* of the gambler after the nth play.

For any $n \geq 1,$ the gambler is assumed to *stop* in case of one of the following conditions:

If $X_n = 0$, he *stops* since he is not allowed to play anymore when he gets broke.
If $X_n = N$, he *stops* due to his decision to stop when he accumulates $\$N$.

Thus,

$$S = \{0, 1, 2 \ldots, N\}$$

will be the *state space* of this special *simple random walk*, where

$$C_1 = \{0\}$$

$$C_2 = \{1, 2, \ldots, N - 1\}$$

$$C_3 = \{N\}$$

$$S = C_1 \cup C_2 \cup C_3$$

It should be noted that, since

$$P_{00} = P_{NN} = 1$$

the states 0 and N are the *absorbing* and the *recurrent* states, while $\{1, 2 \ldots, N - 1\}$ are the *transient* states.

Let $P_k(N)$ be the probability that, starting with $\$k$, the gambler's total fortune will be N. It can be shown that

$$P_k(N) = \begin{cases} \frac{1-((1-p)/p)^k}{1-((1-p)/p)^N} & \text{if } p \neq \frac{1}{2} \\ \frac{k}{N} & \text{if } p = \frac{1}{2} \end{cases}$$

Example 1 Assume that the gambler wins $\$1$ with probability 0.4 at each play. Hence, the probability that the gambler will accumulate $\$25$, starting with $\$10$, will be

$$P_{10}(25) = \frac{1 - (0.6/0.4)^{10}}{1 - (0.6/0.4)^{25}} = 0.002$$

If the gambler wins $\$1$ with probability 0.5 at each play, then the probability that the gambler will accumulate $\$25$, starting with $\$10$, will be

$$P_{10}(25) = \frac{10}{25} = 0.4$$

Remark 2 It can also be shown that the probability for *infinite fortune*, starting with $\$k$, will be

$$P_k(\infty) = \begin{cases} 1 - ((1-p)/p)^k & \text{if } p > \frac{1}{2} \\ 0 & \text{if } p \le \frac{1}{2} \end{cases}$$

Example 1 (revisited 1) If the gambler wins \$1 with probability 0.55 at each play, then the probability for *infinite fortune*, starting with \$10, will be

$$P_{10}(\infty) = 1 - (0.45/0.55)^{10} = 0.8656$$

13.2 Branching Process

Definition 4 (*Branching process*) We consider a *population*, in which each individual is able to *produce i* offsprings with probability P_i *independently* from each other.

Let

X_0 : Number of individuals in the *initial* generation(Size of the *initial* generation)

and generally

$$X_n : \text{Size of the } n\text{th generation}$$

Hence,

$$\{X_n; n \ge 0\}$$

will be a special $DTMC$, called as *branching process*, since the size of each generation depends only on the size of the *previous* generation.

WARNING In a branching process, it is assumed that the individuals produce offsprings without mating. It is also assumed that the individuals in a generation *do not survive* to the next generation, but their offsprings.

Definition 5 (*Expected value and variance of the size of a generation in a branching process*) If we assume that $X_0 = 1$, then it can be shown that the expected value of the size of the nth generation is

$$E[X_n] = \mu^n$$

and the variance of the size of the nth generation is

$$Var(X_n) = \begin{cases} \sigma^2 \mu^{n-1} \left(\frac{1-\mu^n}{1-\mu} \right) & \text{if } \mu \ne 1 \\ n\sigma^2 & \text{if } \mu = 1 \end{cases}$$

Table 13.2 Three conditions for the probability of the eventual population extinction in a branching process

Condition 1 $\mu < 1$	Condition 2 $\mu = 1$	Condition 3 $\mu > 1$
$\pi_0 = 1$	$\pi_0 = 1$	$\pi_0 < 1$ where $\pi_0 = \sum_{i=0}^{\infty} \pi_0^i P_i$

for $n \geq 1$, where

$$\mu = \sum_{i=0}^{\infty} i P_i$$

is the expected value of the number of offsprings of an individual in *any* generation, and σ^2 is the variance of the number of offsprings of an individual in *any* generation.

Definition 6 (*Probability of the eventual population extinction in a branching process*) Let π_0 be the probability of the eventual population extinction in a branching process, given that $X_0 = 1$. One of the three conditions given in Table 13.2 should hold.

Example 2 Assume that there is initially one individual, and this individual produces 0 offspring with probability 0.2, 1 offspring with probability 0.1, 2 offsprings with probability 0.6, and 3 offsprings with probability 0.1. These values are assumed to remain constant for any individual at any generation.

(a) Determine the expected value and the variance of the size of the 8th generation.
(b) What is the probability that the population will eventually become extinct?

Solution

(a) Recall that

μ : Expected value of the number of offsprings of an individual in *any* generation

Hence,

$$\mu = 0P_0 + 1P_1 + 2P_2 + 3P_3$$
$$= 0(0.2) + 1(0.1) + 2(0.6) + 3(0.1) = 1.6$$

and the expected value of the size of the 8th generation will be

$$E[X_8] = (1.6)^8 = 42.95$$

Since

$$\sigma^2 = \left[0^2(0.2) + 1^2(0.1) + 2^2(0.6) + 3^2(0.1)\right] - (1.6)^2 = 0.84$$

and $\mu = 1.6 \neq 1$, the variance of the size of the 8th generation will be

$$Var(X_8) = (0.84)(1.6)^7 \left(\frac{1 - (1.6)^8}{1 - 1.6}\right) = 1{,}576.51$$

(b) Since $\mu = 1.6 > 1$, *Condition* 3 and $\pi_0 < 1$ will hold, and

$$\pi_0 = \pi_0^0 P_0 + \pi_0^1 P_1 + \pi_0^2 P_2 + \pi_0^3 P_3$$
$$\pi_0 = \pi_0^0(0.2) + \pi_0^1(0.1) + \pi_0^2(0.6) + \pi_0^3(0.1)$$
$$0.1\pi_0^3 + 0.6\pi_0^2 - 0.9\pi_0 + 0.2 = 0$$

Note that

$$0.1\pi_0^3 + 0.6\pi_0^2 - 0.9\pi_0 + 0.2 = \left(0.1\pi_0^2 + 0.7\pi_0 - 0.2\right)(\pi_0 - 1)$$

Based on the condition that $\pi_0 < 1$ holds, we should find the roots of $\left(0.1\pi_0^2 + 0.7\pi_0 - 0.2\right)$. When we solve this equation, we find two roots, i.e. 0.2749 and -7.2749. As a result, the probability of the eventual population extinction will be

$$\pi_0 = 0.2749$$

13.3 Hidden Markov Chains

Definition 7 (*Hidden Markov chains—Hidden Markov Models—HMM*) Let

$$\{X_n; n \geq 0\}$$

be a *DTMC* with transition probability matrix P, whose states *cannot be observed*. Let

$$\{Y_n; n \geq 0\}$$

be a discrete-time stochastic process, where each random variable Y_n corresponds to an *observation (emission)* related to each unobserved state X_n such that each observation (emission) is from a finite set Γ. If

$$P\left(Y_n = i_y | X_0, Y_0, X_1, Y_1, X_2, Y_2, \ldots, X_{n-1}, Y_{n-1}, X_n = i_x\right)$$
$$= P\left(Y_n = i_y | X_n = i_x\right) = P_{i_x i_y}$$

holds for all $n \geq 0$, and for all i_x, i_y, where

$$\sum_{i_y \in \Gamma} P_{i_x i_y} = 1 \quad \forall i_x$$

then this Markov chain is called as a *hidden Markov chain*. The matrix Q with entries $P_{i_x i_y}$ is called as an *emission matrix*.

Example 3 We consider a patient who can be in one of two *unobserved states* in any week, i.e. state 1 and state 2. The probability that the patient will be in state 1 next week given that he is in state 1 this week is 0.8, while the probability that he will be in state 2 next week given that he is in state 2 this week is 0.3. The blood test results of the patient are monitored continuously, and the states of the blood test results can be *bad* (*B*), *mediocre* (*M*) or *good* (*G*). Given that he is in state 1, the blood test results can be *bad* with probability 0.1 and *good* with probability 0.5; while given that he is in state 2, the blood test results can be *bad* with probability 0.2 and *good* with probability 0.6. Hence, the transition probability matrix including the *transitions* between the *unobserved states* will be

$$P = \begin{bmatrix} 0.8 & 0.2 \\ 0.7 & 0.3 \end{bmatrix}$$

Considering that $\Gamma = \{B, M, G\}$, the *emission matrix* will be

$$Q = \begin{bmatrix} 0.1 & 0.4 & 0.5 \\ 0.2 & 0.2 & 0.6 \end{bmatrix}$$

This hidden Markov chain model can also be illustrated with the state transition diagram given in Fig. 13.1.

13.4 Time-Reversible Discrete-Time Markov Chains

Definition 8 (*Time-reversible discrete-time Markov chains*) Let

$$\{X_n; n \geq 0\}$$

be a *stationary* and *ergodic* discrete-time Markov chain. Recall the *Markovian property* as

Fig. 13.1 State transition
diagram for Example 3

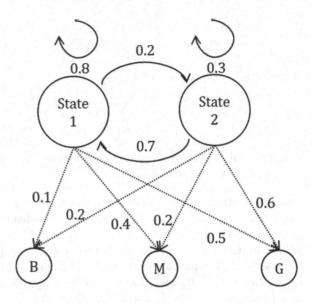

$$P(X_{n+1} = j | X_0 = i_0, X_1 = i_1, X_2 = i_2, \ldots, X_n = i) = P(X_{n+1} = j | X_n = i) = P_{ij}$$

which means "*the future is independent of the past given the present state*".
The following will be a *reversed process*

$$P(X_n = j | X_0 = i_0, X_1 = i_1, X_2 = i_2, \ldots, X_{n+1} = i) = P(X_n = j | X_{n+1} = i) = Q_{ij}$$

and will have the Markovian property by *symmetry* which means "*the present is
independent of the past given the future state*". If

$$P_{ij} = Q_{ij} \quad \text{for all } i, j \in S$$

holds, then this Markov chain is said to be a *time-reversible discrete-time Markov
chain*.

Remark 3 Let π_i, π_j be the limiting probabilities of the states $i, j \in S$. Hence, it can
be shown that

$$Q_{ij} = \frac{\pi_j P_{ji}}{\pi_i} = P_{ij} \quad \text{for all } i, j \in S$$

should hold, and finally

$$\pi_i P_{ij} = \pi_j P_{ji} \quad \text{for all } i, j \in S$$

should be verified for a discrete-time Markov chain to be *time-reversible*.

Example 4 Consider a Markov chain with state space $S = \{0, 1\}$ and with transition probability matrix

$$P = \begin{bmatrix} 1/3 & 2/3 \\ 2/3 & 1/3 \end{bmatrix}$$

By using the balance equations

$$\pi_0 = \frac{1}{3}\pi_0 + \frac{2}{3}\pi_1$$

or

$$\pi_1 = \frac{2}{3}\pi_0 + \frac{1}{3}\pi_1$$

and by using the equation

$$\pi_0 + \pi_1 = 1$$

the limiting probabilities of the states can be computed as

$$\pi_0 = \pi_1 = \frac{1}{2}$$

Since $\pi_0 = \pi_1 = \frac{1}{2}$, by recalling that $Q_{ij} = \frac{\pi_j P_{ji}}{\pi_i}$ for all $i, j \in S$, $Q_{01} - P_{10} = \frac{2}{3}$, $Q_{10} = P_{01} = \frac{2}{3}$, and accordingly

$$Q = \begin{bmatrix} 1/3 & 2/3 \\ 2/3 & 1/3 \end{bmatrix}$$

As a result $P = Q$, and this Markov chain is a time-reversible discrete-time Markov chain.

13.5 Markov Decision Process

Definition 9 (*Markov Decision Process—MDP*) Consider a stochastic process with state space S and *action space* A such that

$$P(X_{n+1} = j | X_0, a_0, \ X_1, a_1, \ldots, X_n = i, a_n = a)$$
$$= P(X_{n+1} = j | X_n = i, a_n = a) = P_{ij}(a)$$

holds for all $n \geq 0$, $i, j \in S$, $a \in A$. It means, the *future state* depends *only* on the *present state*, and the *present action* taken after observing the present state. $P_{ij}(a)$ can be interpreted as "the probability that the process will be in state j *one step later* given that it is in state i now, and action a is taken". Accordingly,

$$\{(X_n, a_n); n \geq 0\}$$

will be a *Markov Decision Process (MDP)*. In an MDP, the rewards can be defined as a stochastic process such that

$$\{R(X_n, a_n); n \geq 0\}$$

where

$R(X_n, a_n)$: Reward earned when the state is X_n and the action taken is a_n

The limiting probabilities for each state i and action a can be defined as a stochastic process such that

$$\{\pi_{ia}; i \in S, a \in A\}$$

where

$$\sum_i \sum_a \pi_{ia} = 1$$

We can use so-called *randomized policy* such that we can define

$\beta_i(a)$: Probability that action a is taken given that MDP is in state i

for all $i \in S$, $a \in A$, where for all $i \in S$, $a \in A$,

$$0 \leq \beta_i(a) \leq 1$$

$$\sum_{a \in A} \beta_i(a) = 1$$

$$\beta_i(a) = \frac{\pi_{ia}}{\sum_{a \in A} \pi_{ia}}$$

Remark 4 As an example of an MDP application, the parameters can be fed into a Linear Programming (LP) model to find the optimal π_{ia} values for the maximization of the total expected reward. For the details of MDP, the reader is referred to the literature provided in the References section.

Example 5 A patient's disease is assumed to be in one of two states. Hence, the *state space* can be defined as

$$S = \{s_1, s_2\}$$

Three different actions are assumed to be available for each state, thus, the *action space* can be defined as

$$A = \{a_{11}, a_{12}, a_{13}, a_{21}, a_{22}, a_{23}\}$$

where a_{11}, a_{12}, a_{13} are the possible actions for state 1, and a_{21}, a_{22}, a_{23} are the possible actions for state 2. *Rewards* can be defined for each state and action as

$$R(s_1, a_{11}) = 10$$

$$R(s_1, a_{12}) = 20$$

$$R(s_1, a_{13}) = -5$$

$$R(s_2, a_{21}) = -7$$

$$R(s_2, a_{22}) = 10$$

$$R(s_2, a_{23}) = 8$$

Note that the rewards can be *negative*, which means that they can be *costs*. The *transition probabilities* can be defined as

$$P(s_1|s_1, a_{11}) = 0.3 \quad P(s_2|s_1, a_{11}) = 0.7$$

$$P(s_1|s_1, a_{12}) = 0.5 \quad P(s_2|s_1, a_{12}) = 0.5$$

$$P(s_1|s_1, a_{13}) = 0.6 \quad P(s_2|s_1, a_{13}) = 0.4$$

$$P(s_1|s_2, a_{21}) = 0.75 \quad P(s_2|s_2, a_{21}) = 0.25$$

$$P(s_1|s_2, a_{22}) = 0.3 \quad P(s_2|s_2, a_{22}) = 0.7$$

$$P(s_1|s_2, a_{23}) = 0.5 \quad P(s_2|s_2, a_{23}) = 0.5$$

The limiting probabilities for each state and action can be defined as

$$\pi_{s_1, a_{11}} = 0.10$$

$$\pi_{s_1, a_{12}} = 0.30$$

$$\pi_{s_1, a_{13}} = 0.05$$

$$\pi_{s_2, a_{21}} = 0.20$$

$$\pi_{s_2, a_{22}} = 0.15$$

$$\pi_{s_2, a_{23}} = 0.20$$

Note that the sum of the limiting probabilities is 1. Based on the limiting probabilities, the following conditional probabilities can be computed for the randomized policy:

$$\beta_{s_1}(a_{11}) = \frac{\pi_{s_1, a_{11}}}{\pi_{s_1, a_{11}} + \pi_{s_1, a_{12}} + \pi_{s_1, a_{13}}} = \frac{0.10}{0.10 + 0.30 + 0.05} = 0.22$$

$$\beta_{s_1}(a_{12}) = \frac{\pi_{s_1, a_{12}}}{\pi_{s_1, a_{11}} + \pi_{s_1, a_{12}} + \pi_{s_1, a_{13}}} = \frac{0.30}{0.10 + 0.30 + 0.05} = 0.67$$

$$\beta_{s_1}(a_{13}) = \frac{\pi_{s_1, a_{13}}}{\pi_{s_1, a_{11}} + \pi_{s_1, a_{12}} + \pi_{s_1, a_{13}}} = \frac{0.05}{0.10 + 0.30 + 0.05} = 0.11$$

$$\beta_{s_2}(a_{21}) = \frac{\pi_{s_2, a_{21}}}{\pi_{s_2, a_{21}} + \pi_{s_2, a_{22}} + \pi_{s_2, a_{23}}} = \frac{0.20}{0.20 + 0.15 + 0.20} = 0.36$$

$$\beta_{s_2}(a_{22}) = \frac{\pi_{s_2, a_{22}}}{\pi_{s_2, a_{21}} + \pi_{s_2, a_{22}} + \pi_{s_2, a_{23}}} = \frac{0.15}{0.20 + 0.15 + 0.20} = 0.28$$

$$\beta_{s_2}(a_{23}) = \frac{\pi_{s_2, a_{23}}}{\pi_{s_2, a_{21}} + \pi_{s_2, a_{22}} + \pi_{s_2, a_{23}}} = \frac{0.20}{0.20 + 0.15 + 0.20} = 0.36$$

Note that

$$\beta_{s_1}(a_{11}) + \beta_{s_1}(a_{12}) + \beta_{s_1}(a_{13}) = 1$$

$$\beta_{s_2}(a_{21}) + \beta_{s_2}(a_{22}) + \beta_{s_2}(a_{23}) = 1$$

Problems

1. Recall the Markov chain with state space $S = \{0, 1, 2\}$ and with transition probability matrix

$$P = \begin{bmatrix} 0.3 & 0.4 & 0.3 \\ 0.5 & 0.2 & 0.3 \\ 0.2 & 0.6 & 0.2 \end{bmatrix}$$

from Example 1 (revisited 2) in Chap. 12. Is this Markov chain a *time-reversible* Markov chain?

Solution Recall from Example 1 (revisited 7) in Chap. 12 that

$$\pi_0 = \frac{23}{66}, \quad \pi_1 = \frac{25}{66}, \quad \pi_2 = \frac{18}{66}$$

Recall also the condition for a time-reversible discrete-time Markov chain

$$\pi_i P_{ij} = \pi_j P_{ji} \quad \text{for all } i, j \in S$$

from Remark 3 in this chapter. As an example,

$$\pi_0 P_{01} = \left(\frac{23}{66}\right)(0.4) = 0.1394$$

$$\pi_1 P_{10} = \left(\frac{25}{66}\right)(0.5) = 0.1894$$

which means $\pi_0 P_{01} \neq \pi_1 P_{10}$. Thus, this Markov chain is *not* a *time-reversible* Markov chain.

Remark 5 It will be sufficient to show for at least one $i, j, \pi_i P_{ij} \neq \pi_j P_{ji}$ to verify that a discrete-time Markov chain is *not* a *time-reversible* discrete-time Markov chain.

2. Consider one fair (F) coin with heads probability 0.5, and one loaded (L) coin with heads probability 0.9. Assume that at each toss, the current coin is switched to the other type with probability 0.3. The initial probabilities are given as $P(F) = 0.4$ and $P(L) = 0.6$.

   (a) Describe this system as a hidden Markov chain.
   (b) Find the joint probability of the states and observations (emissions) given in Table 13.3.

Solution

Table 13.3 States and emissions for problem 2

State	F	L	F	L	L
Emission	H	T	H	T	H

Fig. 13.2 State transition
diagram for problem 2

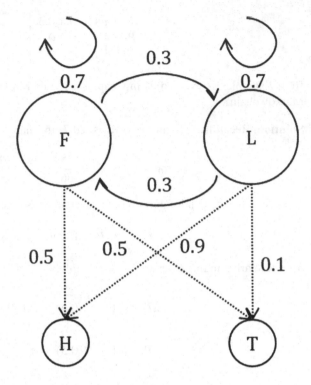

(a) The state transition diagram given in Fig. 13.2 describes the system as a hidden
Markov chain. In this diagram, the *unobserved states* are Fair (*F*) coin and
Loaded (*L*) coin, while the *observations* (*emissions*) are Heads (*H*) and Tails
(*T*).

The transition probability matrix including the *transitions* between the *unobserved
states* will be

$$P = \begin{bmatrix} 0.7 & 0.3 \\ 0.3 & 0.7 \end{bmatrix}$$

and by considering that $\Gamma = \{H, T\}$, the *emission matrix* will be

$$Q = \begin{bmatrix} 0.5 & 0.5 \\ 0.9 & 0.1 \end{bmatrix}$$

(b) Table 13.4 provides the emission probabilities for each state, and the transition
probabilities between the states.

By using the *multiplication rule* from Definition 2 in Chap. 3, the joint probability
of these states and emissions can be computed as

Table 13.4 States, emissions, conditional probabilities for problem 2

State	F	L	F	L	L
Emission	H	T	H	T	H
P(emission|state)	0.5	0.1	0.5	0.1	0.9
P(state|state)	–	0.3	0.3	0.3	0.7

$$P(\{F, H\}, \{L, T\}, \{F, H\}, \{L, T\}, \{L, H\}) = (0.9)(0.7)(0.5)^2(0.4)(0.3)^3(0.1)^2$$
$$= 0.000017$$

by considering the initial probability $P(F) = 0.4$.

3. In a branching process, the number of offsprings each individual produces is *Poisson* distributed having rate λ. Assume $X_0 = 1$, and consider $\lambda = 1$, $\lambda = 3$ for the following questions:

   (a) Determine the expected combined population through the 9th generation.
   (b) What is the probability that the population will eventually become extinct?

Solution

(a) Recall that for $X_0 = 1$,

$$E[X_n] = \mu^n \quad \text{for } n \geq 1$$

Accordingly, the expected combined population through the 9th generation will be

$$E[X_0 + X_1 + X_2 + \cdots + X_9] = E[X_0] + E[X_1] + \cdots + E[X_9]$$
$$= 1 + \mu + \mu^2 + \cdots + \mu^9$$

where

μ : Expected value of the number of offsprings of an individual

Since the number of offsprings each individual produces is *Poisson* distributed having rate λ, $\mu = \lambda$ holds.
For $\mu = \lambda = 1$,

$$E[X_0 + X_1 + X_2 + \cdots + X_9] = 10$$

For $\mu = \lambda = 3$,

$$E[X_0 + X_1 + X_2 + \cdots + X_9] = \frac{1 - \mu^{10}}{1 - \mu} = \frac{1 - (3)^{10}}{1 - 3} = 29,524$$

by the sum of the *finite geometric series*.

(b) For $\mu = \lambda = 1$, *Condition* 2 will be the case for the probability of the eventual population extinction from Table 13.2, and

$$\pi_0 = 1$$

For $\mu = \lambda = 3$, *Condition* 3 will be the case for the probability of the eventual population extinction from Table 13.2, and

$$\pi_0 < 1$$

where

$$\pi_0 = \sum_{i=0}^{\infty} \pi_0^i P_i = \pi_0^0 \left(e^{-\lambda \frac{\lambda^0}{0!}} \right) + \pi_0^1 \left(e^{-\lambda \frac{\lambda^1}{1!}} \right) + \pi_0^2 \left(e^{-\lambda \frac{\lambda^2}{2!}} \right) + \pi_0^3 \left(e^{-\lambda \frac{\lambda^3}{3!}} \right) + \cdots$$

$$= e^{-\lambda} \left(\frac{\lambda^0}{0!} \pi_0^0 + \frac{\lambda^1}{1!} \pi_0^1 + \frac{\lambda^2}{2!} \pi_0^2 + \frac{\lambda^3}{3!} \pi_0^3 + \cdots \right)$$

is the equation to find the probability of the eventual population extinction.
    Since

$$\left(\frac{\lambda^0}{0!} \pi_0^0 + \frac{\lambda^1}{1!} \pi_0^1 + \frac{\lambda^2}{2!} \pi_0^2 + \frac{\lambda^3}{3!} \pi_0^3 + \cdots \right) \cong e^{\lambda \pi_0}$$

by Taylor expansion,

$$\pi_0 \cong e^{-\lambda} e^{\lambda \pi_0} = e^{\lambda(\pi_0 - 1)}$$

will be obtained, and for $\mu = \lambda = 3$ it will be equal to

$$\pi_0 \cong e^{3(\pi_0 - 1)}$$

which can be solved numerically as $\pi_0 \cong 0.99999$, since $\pi_0 < 1$ should hold.

4. Consider a Markov chain $\{X_n; n \geq 0\}$ with *restricted state space* $S = \{1, 2, 3, 4, 5\}$.

   (a) Define this Markov chain as a *simple symmetric random walk*.
   (b) Build the state transition diagram for this simple symmetric random walk.
   (c) Build the transition probability matrix for this simple symmetric random walk.

Solution

(a) $\{X_n; n \geq 0\}$ can be defined as a simple symmetric random walk with *restricted state space*

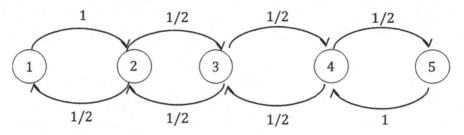

Fig. 13.3 State transition diagram for problem 4

$$S = \{1, 2, 3, 4, 5\}$$

and with following transition probabilities

$$P_{i,\,i+1} = P_{i,\,i-1} = \frac{1}{2} \quad \text{for } 2 \le i \le 4$$

$$P_{i,i+1} = 1 \quad \text{for } i = 1$$

$$P_{i,i-1} = 1 \quad \text{for } i = 5$$

(b) The state transition diagram for this simple symmetric random walk can be illustrated as in Fig. 13.3.

(c) The transition probability matrix for this simple symmetric random walk will be

$$P = \begin{bmatrix} 0 & 1 & 0 & 0 & 0 \\ 1/2 & 0 & 1/2 & 0 & 0 \\ 0 & 1/2 & 0 & 1/2 & 0 \\ 0 & 0 & 1/2 & 0 & 1/2 \\ 0 & 0 & 0 & 1 & 0 \end{bmatrix}$$

Chapter 14
Continuous-Time Markov Chains

Abstract The first part of this chapter presented the definition of a continuous-time Markov chain with two properties, and the introduction of a *B&D* process with some special examples such as homogeneous Poisson process as a pure birth process, and the population model as a *B&D* process. The second part of the chapter was an extension of a *B&D* process to the *B&D* queueing models including the general *B&D* queueing model, *M/M/*1 queueing model, and *M/M/m* queueing model. Finally, the last part of the chapter introduced Kolmogorov's backward/forward equations, the infinitesimal generator matrix of a *CTMC*, and the time-reversibility condition for a *CTMC*. Several examples and problems including those related to the population model, the server systems, and the patient states have been provided and solved.

After providing the basic definitions and the basic properties for the continuous-time Markov chains in Chap. 12, this chapter will introduce two distinctive properties of a *CTMC*, and the basics of the birth & death (*B&D*) process as a special and very important type of *CTMC*. The basic properties of the *B&D* queueing models including the general *B&D* queueing model, *M/M/*1 queueing model, and *M/M/m* queueing model will be characterized as a special type of *B&D* process, and these models will be illustrated with some examples and problems. *B&D* queueing models provided in this chapter will be strong background for the queueing models in Chap. 15, and the details of the *B&D* queueing models including the balance equations will be left to Chap. 15. Since the time spent in a state is continuous in a *CTMC*, to be able to compute the transition probabilities, Kolmogorov's backward/forward equations are required, which will be defined in the last sub-section, and illustrated with some examples and one problem. In this sub-section, the infinitesimal generator matrix of a *CTMC*, and the time-reversibility condition of a *CTMC* will also be defined.

© Springer Nature Switzerland AG 2019
E. Bas, *Basics of Probability and Stochastic Processes*,
https://doi.org/10.1007/978-3-030-32323-3_14

14.1 Continuous-Time Markov Chain and Birth & Death Process

Definition 1 (*Continuous-time Markov chain—CTMC*) A continuous-time Markov chain can be defined with the properties given in Table 14.1.

Example 1 Consider Problem 1 in Chap. 12. Recall that S = {0, 1, 2, 3, 4}, where state 0 is freedom. If we assume that the rat spends *exponentially* distributed time in each transient state with mean 6 min, then

$$\{X(t); t \geq 0\}$$

will be a *CTMC*, where

$$X(t) : \text{The cell of the rat at time } t,$$
$$v_i = \frac{60}{6} = 10 \, \text{per hour}, \quad i = 1, 2, 3, 4$$
$$v_0 = 0$$

Note that the transition probability matrix will be the same as in *DTMC* such that

$$P = \begin{bmatrix} 1 & 0 & 0 & 0 & 0 \\ 0 & 0 & 1/2 & 1/2 & 0 \\ 0 & 1/2 & 0 & 0 & 1/2 \\ 0 & 1/2 & 0 & 0 & 1/2 \\ 1/3 & 0 & 1/3 & 1/3 & 0 \end{bmatrix}$$

Remark 1 Note that state 0 is an absorbing state since $v_0 = 0$. Any state i of a *CTMC* is an absorbing state if $v_i = 0$.

Definition 2 (*Birth & death—B&D—process*) A birth & death (*B&D*) process is a special *CTMC*. At each state $i \in S$, it is only possible to make a transition into

$$\text{state } i + 1 \text{ with } birth \, rate \, \lambda_i$$

or into

Table 14.1 Two basic properties of a continuous-time Markov chain

Property 1	Property 2
Given that a *CTMC* is in state i, it spends in that state *exponentially* distributed time Z_i with rate v_i, that is $Z_i \sim exp(v_i)$	Whenever a *CTMC* makes a transition, the probability of transition from state i to any other state j will be P_{ij}, where $\sum_{j \in S, j \neq i} P_{ij} = 1$

state $i - 1$ with *death rate* μ_i

Accordingly,

$$v_i = \lambda_i + \mu_i \quad i \in S$$

Remark 2 Please note that for a *B&D* process, $Z_i = \min(X_i, Y_i)$ holds, where

Z_i : Time from state i to the next transition (to the next birth or next death)
X_i : Time from state i to state $i + 1$ (to the *next birth*)
Y_i : Time from state i to state $i - 1$ (to the *next death*)

Since

$$X_i \sim exp(\lambda_i) \quad i \in S$$
$$Y_i \sim exp(\mu_i) \quad i \in S$$

it follows that

$$Z_i \sim exp(\lambda_i + \mu_i) \quad i \in S$$

by *further property* 1 of exponential random variable from Remark 3 in Chap. 9, which means

$$v_i = \lambda_i + \mu_i \quad i \in S$$

Remark 3 Given that a *CTMC* is in state $i \in S$ now, the probability that it makes a transition into state $i + 1$ will be

$$P_{i,i+1} = P(X_i < Y_i) = \frac{\lambda_i}{\lambda_i + \mu_i} \quad i \in S$$

and the probability that it makes a transition into state $i - 1$ will be

$$P_{i,i-1} = P(Y_i < X_i) = \frac{\mu_i}{\lambda_i + \mu_i} \quad i \in S$$

by *further property* 2 of exponential random variable from Remark 3 in Chap. 9.

Definition 3 (*Homogeneous Poisson Process as a pure birth process*) A homogeneous Poisson process with rate λ can be characterized as a *pure birth process*

$$\{X(t); t \geq 0\}$$

where

Table 14.2 Two basic properties of a homogeneous Poisson process as a pure birth process

Property 1	Property 2
A homogeneous Poisson process with rate λ spends in state i *exponentially* distributed time with rate $$v_i = \lambda_i + \mu_i = \lambda_i = \lambda \quad i \geq 0$$	The probability of transition from state i to state $i + 1$ will be $$P_{i,i+1} = 1 \quad i \geq 0$$ The probability of transition from state i to any other state $j \neq i + 1$ will be $$P_{i,j} = 0 \quad j \neq i + 1$$

$X(t)$ Total number of events at time t.

Two basic properties given in Table 14.2 characterize a homogeneous Poisson Process as a pure birth process.

Remark 4 Note that the total number of events for a Poisson process is always *nondecreasing* with time, which means, there is *no death*, and $\mu_i = 0$ holds for all $i \geq 0$.

Definition 4 (*Population model as a B&D process*) A population model can be characterized as a *B&D* process

$$\{X(t); t \geq 0\}$$

where

$X(t)$ Total number of individuals in the population at time t

Two basic properties given in Table 14.3 characterize a population model as a *B&D* process.

Remark 5 When there are i individuals in the population, the time to the *first birth* is an *exponential* random variable with rate $\lambda_i = i\lambda$, where each individual is assumed to give birth to one offspring with an exponential rate λ, and the time to the *first*

Table 14.3 Two basic properties of a population model as a *B&D* process

Property 1	Property 2
Given that there are i individuals in the population, the population stays in that state *exponentially* distributed time with rate $$v_i = \lambda_i + \mu_i \quad i > 0$$ where $$\lambda_i = i\lambda \quad i > 0$$ is the *birth rate* for state i, and $$\mu_i = i\mu \quad i > 0$$ is the *death rate* for state i.	The probability of transition from state i to state $i + 1$, i.e. the probability of transition to the *next birth* when there are i individuals in the population will be $$P_{i,i+1} = \frac{\lambda_i}{\lambda_i + \mu_i} = \frac{\lambda}{\lambda + \mu} \quad i > 0$$ The probability of transition from state i to state $i - 1$, i.e. the probability of transition to the *next death* when there are i individuals in the population will be $$P_{i,i-1} = \frac{\mu_i}{\lambda_i + \mu_i} = \frac{\mu}{\lambda + \mu} \quad i > 0$$

death is an *exponential* random variable with rate $\mu_i = i\mu$, where each individual is assumed to die with an exponential rate μ. The basic assumption in this model is that every individual can give birth to one offspring without mating, and there is no immigration to the population, thus, $\lambda_0 = 0$.

14.2 Birth & Death Queueing Models

Definition 5 (*A general B&D queueing model with infinite capacity*) A general *B&D* queueing model can be characterized as a *B&D* process,

$$\{X(t); t \geq 0\}$$

where

$X(t)$ Total number of objects in the system at time t

Note that
λ_i: *Arrival* rate (birth rate)
μ_i: *Departure* rate (death rate)
when there are i objects in the system.

Two basic properties given in Table 14.4 characterize a general *B&D* queueing model with infinite capacity.

Definition 6 (*M/M/1 queueing model with infinite capacity*) An *M/M/1* queueing model can be characterized as a special *B&D* queueing model,

$$\{X(t); t \geq 0\}$$

Table 14.4 Two basic properties of a general *B&D* queueing model with infinite capacity

Property 1	Property 2
Given that there are i objects in the system, the system will stay in that state *exponentially* distributed time with rate $v_i = \lambda_i + \mu_i \quad i \geq 0$ Note that $v_0 = \lambda_0$ since $\mu_0 = 0$	The probability of transition from state i to state $i + 1$, i.e. the probability of transition to the *next arrival* when there are i objects in the system will be $P_{i,i+1} = \frac{\lambda_i}{\lambda_i + \mu_i} \quad i \geq 0$ The probability of transition from state i to state $i - 1$, i.e. the probability of transition to the *next departure* when there are i objects in the system will be $P_{i,i-1} = \frac{\mu_i}{\lambda_i + \mu_i} \quad i > 0$ Note that $P_{01} = 1$

Table 14.5 Two basic properties of an $M/M/1$ queueing model with infinite capacity

Property 1	Property 2
Given that there are i objects in the system, the system will stay in that state *exponentially* distributed time with rate $v_i = \lambda_i + \mu_i \quad i \geq 0$ where $\lambda_i = \lambda \quad i \geq 0$ $\mu_i = \mu \quad i > 0$ Finally, $v_i = \lambda + \mu \quad i > 0$ Note that $v_0 = \lambda$ since $\mu_0 = 0$	The probability of transition from state i to state $i + 1$, i.e. the probability of transition to the *next arrival* when there are i objects in the system will be $P_{i,i+1} = \frac{\lambda}{\lambda+\mu} \quad i > 0$ The probability of transition from state i to state $i - 1$, i.e. the probability of transition to the *next departure* when there are i objects in the system will be $P_{i,i-1} = \frac{\mu}{\lambda+\mu} \quad i > 0$ Note that $P_{01} = 1$

where

$X(t)$ Total number of objects in the system at time t

Two basic properties given in Table 14.5 characterize an $M/M/1$ queueing model with infinite capacity.

Remark 6 In an $M/M/1$ queueing model, the first M refers to *Markovian*, which means the arrivals of the objects are according to a *Poisson process*, hence the inter-arrival times are *exponential* random variables, and *Markovian* due to the unique *memoryless property*; the second M refers to *Markovian*, which means the service times are *exponential* random variables, and *Markovian*, and 1 refers to the number of servers.

Example 2 We consider a system with 1 server, whose service time is *exponentially* distributed with mean 3 min. Customers arrive to this system according to a *Poisson process* at a rate of 15 per hour. Thus, this system is an $M/M/1$ queueing system. Accordingly, the time until the next arrival *or* departure will be

$$Z = \min(X, Y)$$

where

X: Time until the next *arrival*
Y: Time until the next *departure*

Since mean service time is 3 min, the service rate (departure rate) will be $\frac{60}{3} = 20$ per hour, and it follows that

$$X \sim exp(15)$$
$$Y \sim exp(20)$$
$$Z \sim \exp(15 + 20) \text{ and } Z \sim exp(35)$$

Accordingly, the expected time until the next arrival *or* departure will be

$$E[Z] = \frac{1}{35}h \cong 1.71\text{min}$$

When there are i customers in the system, the probability of the *next arrival* will be

$$P_{i,i+1} = \frac{15}{35} = \frac{3}{7} \quad i > 0$$

and the probability of the *next departure* will be

$$P_{i,i-1} = \frac{20}{35} = \frac{4}{7} \quad i > 0$$

Note that $P_{01} = 1$.

Definition 7 (*M/M/m queueing model with infinite capacity*) An *M/M/m* queueing model can be characterized as a special *B&D* queueing model,

$$\{X(t); t \geq 0\}$$

where

$X(t)$ Total number of objects in the system at time t

Two basic properties given in Table 14.6 characterize an *M/M/m* queueing model with infinite capacity.

Remark 7 In an *M/M/m* queueing model with infinite capacity, $0 \leq i \leq m$ is the case of no objects waiting in queue. Thus, the time to the first departure will be an *exponential* random variable with rate $\mu_i = i\mu$ by *further property* 1 of exponential random variable, where i refers to the number of the objects in the system. On the

Table 14.6 Two basic properties of an $M/M/m$ queueing model with infinite capacity

Property 1	Property 2
Given that there are i objects in the system, the system will stay in that state *exponentially* distributed time with rate $v_i = \lambda_i + \mu_i$ where $\lambda_i = \lambda \quad i \geq 0$ $\mu_i = \begin{cases} i\mu & 0 \leq i \leq m \\ m\mu & i > m \end{cases}$	The probability of transition from state i to state $i + 1$, i.e. the probability of transition to the *next arrival* when there are i objects in the system will be $P_{i,i+1} = \frac{\lambda}{\lambda+\mu_i} \quad i \geq 0$ The probability of transition from state i to state $i - 1$, i.e. the probability of transition to the *next departure* when there are i objects in the system will be $P_{i,i-1} = \frac{\mu_i}{\lambda+\mu_i} \quad i > 0$

other hand, $i > m$ is the case of all servers busy and some objects waiting in queue. Thus, the time to the first departure will be an *exponential* random variable with rate $\mu_i = m\mu$ by *further property* 1 of exponential random variable, where m refers to the number of customers being served, which means the number of all servers. Note that $P_{01} = 1$ holds.

Example 3 Consider a system with 4 servers. Each server's service time is *exponentially* distributed with rate 15 per hour. Customers arrive to this system according to a *Poisson process* at a rate of 10 per hour. Thus, this system is an *M/M/4* queueing system. When there are i customers in the system, the *arrival rates* will be

$$\lambda_i = \lambda = 10 \quad i \geq 0$$

and the *departure rates* will be

$$\mu_i = \begin{cases} 15i & 0 \leq i \leq 4 \\ 60 & i > 4 \end{cases}$$

Accordingly, the probability of the *next arrival* will be

$$P_{i,i+1} = \frac{\lambda}{\lambda + \mu_i} = \frac{10}{10 + 15i} \quad 0 \leq i \leq 4$$

$$P_{i,i+1} = \frac{\lambda}{\lambda + \mu_i} = \frac{10}{10 + 60} = \frac{1}{7} \quad i > 4$$

and the probability of the *next departure* will be

$$P_{i,i-1} = \frac{\mu_i}{\lambda + \mu_i} = \frac{15i}{10 + 15i} \quad 1 \leq i \leq 4$$

$$P_{i,i-1} = \frac{\mu_i}{\lambda + \mu_i} = \frac{60}{10 + 60} = \frac{6}{7} \quad i > 4$$

14.3 Kolmogorov's Backward/Forward Equations, Infinitesimal Generator Matrix of a *CTMC* and Time-Reversibility of a *CTMC*

Definition 8 (*Transition probability function and Kolmogorov's backward/forward equations*) Recall from Chap.12 that

$$P(X(t + s) = j | X(s) = i) = P_{ij}(t) \quad \text{for} \quad s \geq 0, t > 0$$

Table 14.7 Kolmogorov's backward and forward equations

Kolmogorov's backward equations	Kolmogorov's forward equations
$P'_{ij}(t) = \sum_{k \neq i} q_{ik} P_{kj}(t) - v_i P_{ij}(t) \quad \forall i, j$	$P'_{ij}(t) = \sum_{k \neq j} q_{kj} P_{ik}(t) - v_j P_{ij}(t) \quad \forall i, j$
are called as *Kolmogorov's backward equations*, where	are called as *Kolmogorov's forward equations*, where
$q_{ik} = v_i P_{ik}$	$q_{kj} = v_k P_{kj}$
is the *instantaneous transition rate* for transition from state i to state k.	is the *instantaneous transition rate* for transition from state k to state j.

is the transition probability function which describes the probability that a *CTMC* will be in *state j t time units later* given that it is in *state i* at *any* time $s \geq 0$. In Table 14.7, Kolmogorov's backward and forward equations are defined which are used to find the transition probability functions of a *CTMC*.

Remark 8 The *instantaneous transition rate* q_{ik} is different from the transition rate v_i such that q_{ik} refers to the transition rate for transition from state i to state k, whereas v_i refers to the transition rate for transition from state i to *any other state*. The similar explanation can also be made for q_{kj}.

Example 4 Kolmogorov's *backward* equations for a *B&D* process will be

$$P'_{ij}(t) = \sum_{k \neq i} q_{ik} P_{kj}(t) - v_i P_{ij}(t) \quad \forall i, j$$
$$= q_{i,i+1} P_{i+1,j}(t) + q_{i,i-1} P_{i-1,j}(t) - v_i P_{ij}(t)$$
$$= \lambda_i P_{i+1,j}(t) + \mu_i P_{i-1,j}(t) - (\lambda_i + \mu_i) P_{ij}(t)$$

for $i > 0$, since

$$v_i = \lambda_i + \mu_i$$
$$q_{i,i+1} = v_i P_{i,i+1} = (\lambda_i + \mu_i) \frac{\lambda_i}{\lambda_i + \mu_i} = \lambda_i$$

and

$$q_{i,i-1} = v_i P_{i,i-1} = (\lambda_i + \mu_i) \frac{\mu_i}{\lambda_i + \mu_i} = \mu_i$$

Note that for $i = 0$,

$$P'_{0j}(t) = \lambda_0 P_{1j}(t) - \lambda_0 P_{0j}(t) = \lambda_0 \big(P_{1j}(t) - P_{0j}(t) \big)$$

Example 5 Kolmogorov's *forward* equations for a *B&D* process will be

$$P'_{ij}(t) = \sum_{k \neq j} q_{kj} P_{ik}(t) - v_j P_{ij}(t) \quad \forall i, j$$

$$= q_{j-1,j} P_{i,j-1}(t) + q_{j+1,j} P_{i,j+1}(t) - v_j P_{ij}(t)$$

$$= \lambda_{j-1} P_{i,j-1}(t) + \mu_{j+1} P_{i,j+1}(t) - (\lambda_j + \mu_j) P_{ij}(t)$$

for $j > 0$, since

$$v_j = \lambda_j + \mu_j$$

$$q_{j-1,j} = v_{j-1} P_{j-1,j} = (\lambda_{j-1} + \mu_{j-1}) \frac{\lambda_{j-1}}{\lambda_{j-1} + \mu_{j-1}} = \lambda_{j-1}$$

and

$$q_{j+1,j} = v_{j+1} P_{j+1,j} = (\lambda_{j+1} + \mu_{j+1}) \frac{\mu_{j+1}}{\lambda_{j+1} + \mu_{j+1}} = \mu_{j+1}$$

Note that for $j = 0$,

$$P'_{i0}(t) = \mu_1 P_{i1}(t) - \lambda_0 P_{i0}(t)$$

Definition 9 (*Infinitesimal generator matrix of a CTMC*) The infinitesimal generator matrix Q of a *CTMC* with state space $S = \{0, 1, 2, \ldots, K\}$ is a *matrix* with entries q_{ij} such that

$$\sum_{j \in S, j \neq i} q_{ij} = -q_{ii} \quad \forall i \in S$$

Note that the sum of each row of an infinitesimal generator matrix of a *CTMC* is *zero*.

Example 6 We consider a two-state *CTMC* with state space $S = \{0, 1\}$. The transition rate at state 0 is λ, while the transition rate at state 1 is μ. Hence, the infinitesimal generator matrix Q of this *CTMC* will be

$$Q = \begin{bmatrix} -\lambda & \lambda \\ \mu & -\mu \end{bmatrix}$$

since

$$q_{01} = v_0 P_{01} = \lambda.1 = \lambda$$

$$q_{10} = v_1 P_{10} = \mu.1 = \mu$$

Definition 10 (*Infinitesimal generator matrix of a B&D process*) The infinitesimal generator matrix Q of a *B&D* process with state space $S = \{0, 1, 2, \ldots K\}$ will be

$$Q = \begin{bmatrix} -\lambda_0 & \lambda_0 & 0 & 0 & \cdots \\ \mu_1 & -(\lambda_1 + \mu_1) & \lambda_1 & 0 & \cdots \\ 0 & \mu_2 & -(\lambda_2 + \mu_2) & \lambda_2 & \cdots \\ 0 & 0 & \mu_3 & -(\lambda_3 + \mu_3) & \cdots \\ \cdots & \cdots & \cdots & \cdots & \cdots \end{bmatrix}$$

As an example,

$$q_{10} = v_1 P_{10} = (\lambda_1 + \mu_1) \frac{\mu_1}{\lambda_1 + \mu_1} = \mu_1$$

$$q_{12} = v_1 P_{12} = (\lambda_1 + \mu_1) \frac{\lambda_1}{\lambda_1 + \mu_1} = \lambda_1$$

and accordingly,

$$q_{11} = -(q_{10} + q_{12}) = -(\lambda_1 + \mu_1)$$

will hold for the 2nd row.

Definition 11 (*Time-reversibility of a CTMC*) Recall that a *DTMC* is time-reversible if

$$\pi_i P_{ij} = \pi_j P_{ji} \quad \text{for all} \ \ i, j \in S$$

holds from Remark 3 in Chap. 13. It can be shown that a *CTMC* is time-reversible if

$$P_i q_{ij} = P_j q_{ji} \quad \text{for all} \ \ i, j \in S$$

holds, where
  P_i, P_j: *Limiting probabilities* of the states i, j for a *CTMC*, respectively
  as defined in Table 12.8 in Chap. 12, and
  q_{ij}, q_{ji}: *Instantaneous transition rates* for transition from state i to state j, and from state j to state i, respectively
  as defined in Table 14.7 in this chapter.

Problems

1. Consider a region of a country in which each individual is assumed to give birth to a baby at an exponential rate of $\frac{1}{3}$ per year, and every individual is assumed to die at an exponential rate of $\frac{1}{65}$ per year. There is also immigration to this region with an exponential rate of $\frac{1}{6}$ per year, and each individual is assumed to leave this region with an exponential rate of $\frac{1}{10}$ per year.

(a) Find the probability that the population will *increase* by 1, when there are *i* individuals.

(b) Find the probability that the population will *decrease* by 1, when there are *i* individuals.

(c) What is the expected time until the *next change* in the population, when there are *i* individuals?

Solution

Let

X_i : Time until the *next increase* in the population, when there are *i* individuals

Y_i : Time until the *next decrease* in the population, when there are *i* individuals

Z_i : Time until the *next change* in the population, when there are *i* individuals

Note that

$$Z_i = min(X_i, Y_i)$$

$$X_i \sim exp\left(\frac{1}{3}i + \frac{1}{6}\right)$$

$$Y_i \sim exp\left(\frac{1}{65}i + \frac{1}{10}i\right) \text{ and } Y_i \sim exp\left(\frac{3}{26}i\right)$$

which means

$$Z_i \sim exp\left(\frac{1}{3}i + \frac{1}{6} + \frac{3}{26}i\right) \text{ and } Z_i \sim exp\left(\frac{35i + 13}{78}\right)$$

(a) The probability that the population will *increase* by 1, when there are *i* individuals will be

$$P_{i,i+1} = P(X_i < Y_i) = \frac{\left(\frac{1}{3}i + \frac{1}{6}\right)}{\left(\frac{35i+13}{78}\right)} = \frac{26i + 13}{35i + 13} \quad i \geq 0$$

(b) The probability that the population will *decrease* by 1, when there are *i* individuals will be

$$P_{i,i-1} = P(Y_i < X_i) = \frac{\left(\frac{3}{26}i\right)}{\left(\frac{35i+13}{78}\right)} = \frac{9i}{35i + 13} \quad i > 0$$

(c) Since

$$Z_i \sim exp\left(\frac{35i + 13}{78}\right)$$

the *expected time* until the next change in the population, when there are i individuals will be

$$E[Z_i] = \frac{1}{\left(\frac{35i+13}{78}\right)} = \frac{78}{35i + 13} \quad i \geq 0$$

2. Consider an $M/M/3$ system in which an arriving customer who sees 5 customers in the system cannot enter the system. Customers arrive to this system, on average, every 10 min, and the average service time is 6 min for each server.

(a) What is the entering rate of customers?
(b) What is the departure rate of customers when all servers are busy?
(c) When there are 2 customers in the system, what is the probability that the next arrival will be *before* the next departure? When there are 2 customers in the system, what is the expected time until the next arrival *or* departure?

Solution
Note that

$$\text{Arrival rate } \lambda = \frac{60}{10} = 6 \text{ per hour}$$
$$\text{Service rate } \mu = \frac{60}{6} = 10 \text{ per hour per server}$$

(a) Let
λ_e: Entering rate of customers

Accordingly,

$$\lambda_e = \lambda(P_0 + P_1 + P_2 + P_3 + P_4) = \lambda(1 - P_5)$$
$$= 6(P_0 + P_1 + P_2 + P_3 + P_4) = 6(1 - P_5)$$

where

P_i: Limiting probability that there are i customers in the system, $i = 0, 1, 2, 3, 4, 5$

(b) The departure rate of customers will be

$$\mu_i = 3\mu = 3(10) = 30 \quad \text{for} \quad 3 \leq i \leq 5$$

when all servers are busy.

(c) Let

> X_2 : Time until the *next arrival*, when there are 2 customers in the system
>
> Y_2 : Time until the *next departure*, when there are 2 customers in the system
>
> Z_2 : Time until the *next arrival or next departure*, when there are
>
> 2 customers in the system

The probability that the next arrival will be *before* the next departure is

$$P(X_2 < Y_2) = \frac{\lambda_2}{\lambda_2 + \mu_2} = \frac{6}{6 + 20} = \frac{3}{13}$$

since

$$\lambda_2 = \lambda = 6$$
$$\mu_2 = 2\mu = 2(10) = 20$$

Since

$$Z_2 = min(X_2, Y_2)$$
$$Z_2 \sim exp(\lambda_2 + \mu_2) \text{ and } Z_2 \sim exp(26)$$

the expected time until the next arrival *or* departure, when there are 2 customers in the system will be

$$E[Z_2] = \frac{1}{26}h \cong 2.31\,min$$

3. A patient can be in two states, i.e. state 1 and state 2. While in state 1, he spends exponentially distributed time with rate λ; while in state 2, he spends exponentially distributed time with rate μ. Write down Kolmogorov's *backward* and *forward* equations for this *B&D* process.

Solution
The patient's transition between the states can be modelled as a *B&D* process such that

$$\lambda_1 = \lambda$$
$$\mu_2 = \mu$$

Recall Kolmogorov's *backward* equations for a *B&D* process from Example 4 such that

$$P'_{ij}(t) = \lambda_i P_{i+1,j}(t) + \mu_i P_{i-1,j}(t) - (\lambda_i + \mu_i) P_{ij}(t) \quad \text{for } i > 0$$

Table 14.8 Kolmogorov's backward and forward equations for problem 3

Kolmogorov's backward equations	Kolmogorov's forward equations
$P'_{11}(t) = \lambda(P_{21}(t) - P_{11}(t))$	$P'_{11}(t) = \mu P_{12}(t) - \lambda P_{11}(t)$
$P'_{12}(t) = \lambda(P_{22}(t) - P_{12}(t))$	$P'_{12}(t) = \lambda P_{11}(t) - \mu P_{12}(t)$
$P'_{21}(t) = \mu(P_{11}(t) - P_{21}(t))$	$P'_{21}(t) = \mu P_{22}(t) - \lambda P_{21}(t)$
$P'_{22}(t) = \mu(P_{12}(t) - P_{22}(t))$	$P'_{22}(t) = \lambda P_{21}(t) - \mu P_{22}(t)$

$$P'_{0j}(t) = \lambda_0\big(P_{1j}(t) - P_{0j}(t)\big) \quad \text{for } i = 0$$

Recall also Kolmogorov's *forward* equations for a *B&D* process from Example 5 such that

$$P'_{ij}(t) = \lambda_{j-1}P_{i,j-1}(t) + \mu_{j+1}P_{i,j+1}(t) - \big(\lambda_j + \mu_j\big)P_{ij}(t) \quad \text{for } j > 0$$
$$P'_{i0}(t) = \mu_1 P_{i1}(t) - \lambda_0 P_{i0}(t) \quad \text{for } j = 0$$

Accordingly, Kolmogorov's *backward* and *forward* equations for the problem of a patient with two states will be as in Table 14.8.

As an example for Kolmogorov's *backward* equations,

$$P'_{11}(t) = \lambda_1 P_{21}(t) + \mu_1 P_{01}(t) - (\lambda_1 + \mu_1)P_{11}(t)$$
$$= \lambda P_{21}(t) - \lambda P_{11}(t) = \lambda(P_{21}(t) - P_{11}(t))$$

will be since

$$\lambda_1 = \lambda$$
$$\mu_1 = 0$$

As an example for Kolmogorov's *forward* equations,

$$P'_{11}(t) = \lambda_0 P_{10}(t) + \mu_2 P_{12}(t) - (\lambda_1 + \mu_1)P_{11}(t)$$
$$= \mu P_{12}(t) - \lambda P_{11}(t)$$

will be since

$$\lambda_0 = 0$$
$$\lambda_1 = \lambda$$
$$\mu_1 = 0$$
$$\mu_2 = \mu$$

Chapter 15
An Introduction to Queueing Models

Abstract This chapter was an introduction to the queueing models. The first part of the chapter introduced very basic concepts including Little's law, Poisson Arrivals See Time Averages, and the Kendall's notation for the classification of the queueing models. In the second part of the chapter, the balance equations for a general $B\&D$ queueing model, and the balance equations and Little's law equations for an $M/M/1$ queueing model with infinite capacity and finite capacity have been provided. The adaptation of the basic formulations to the cases with more than one server, with varying arrival rates and departure rates, and with finite capacity has been illustrated with some examples and problems including the ones related to the arrivals of the customers, and the arrivals of the machines to a repair system.

After providing the essential background related to the basics of Markov chains, $CTMC$, the $B\&D$ processes, and the $B\&D$ queueing models in Chaps. 12 and 14; this chapter will introduce some basic definitions for queueing models such as Little's law, balance equations for a general $B\&D$ queuing model, and balance equations and Little's law equations for an $M/M/1$ queueing model with infinite capacity and finite capacity. Although Little's law equations will be provided as formulations for an $M/M/1$ queueing model, the examples and problems will show how Little's law equations can be derived for other $B\&D$ queueing models. In the problems section, a problem related to a one-server system with bulk service will also be presented as a special case of serving more than one customer at a time.

15.1 Basic Definitions

Definition 1 (*Little's law*) Little's law is defined as given in Table 15.1.

Remark 1 Little's law is also called as *conservation equation*. λ_a can be different from the *arrival rate* due to the system capacity or due to the variable arrival rates. Note that arrivals *do not* have to be according to a *Poisson process*.

Example 1 Customers arrive to a bank with a rate of 5 per hour. A customer spends, on average, 8 min in the bank; while the average waiting time in queue is 3 min per

© Springer Nature Switzerland AG 2019
E. Bas, *Basics of Probability and Stochastic Processes*,
https://doi.org/10.1007/978-3-030-32323-3_15

Table 15.1 Little's law

Number of objects in the system	Number of objects waiting in queue
$L = \lambda_a W$ is *Little's Law*, where L : Average number of objects in the system λ_a : Average arrival rate of objects W : Average amount of time an object spends in the system	$L_Q = \lambda_a W_Q$ is a variation of *Little's Law*, where L_Q : Average number of objects waiting in queue λ_a : Average arrival rate of objects W_Q : Average amount of time an object spends waiting in queue

customer. Hence, the average number of customers in the bank will be

$$L = \lambda_a W = \lambda W$$
$$= 5\left(\frac{8}{60}\right) = \frac{2}{3}$$

and the average number of customers waiting in queue will be

$$L_Q = \lambda_a W_Q = \lambda W_Q$$
$$= 5\left(\frac{3}{60}\right) = \frac{1}{4}$$

Suppose that whenever there are 2 customers in the bank, a new customer *cannot enter* the bank. Accordingly, the average arrival rate of customers (*entering rate* of customers) will be

$$\lambda_a = \lambda(P_0 + P_1) = \lambda(1 - P_2) = 5(P_0 + P_1) = 5(1 - P_2)$$

where

P_i: Limiting probability that there are i customers in the bank, $i = 0, 1, 2$

Definition 2 (*Poisson Arrivals See Time Averages—PASTA*) If objects arrive to a queueing system according to a *Poisson process*, then

$$P_i = a_i$$

holds, where

P_i Limiting probability that there are i objects in the system
a_i Proportion of arrivals finding i objects in the system

Remark 2 In words, *PASTA* principle means that the fraction of arrivals finding i objects in the system is exactly the same as the fraction of the time the system includes i objects. *PASTA* principle is also called as *Random Observer Property (ROP)*.

Definition 3 (*Kendall's notation for the classification of the queueing models*)
Kendall's notation for the classification of the queueing models can be shown as

$$A/B/m/K/n/Q$$

where

A : Distribution of interarrival times
B : Distribution of service times
m : Number of parallel servers
K : Capacity of the queueing system including the one(s) being served
n : Population size
Q : Service discipline (Queue discipline)

Example 2

$$M/M/1/\text{infinity}/\text{infinity}/FCFS$$

is the Kendall's notation for an *M/M/*1 queueing model with infinite capacity, infinite
population size, and *First-Come First-Served* (*FCFS*) service (queue) discipline. As
explained in Remark 6 of Chap. 14, M refers to Markovian. Some other notations
for distributions are D for deterministic, G for general distribution, *GI* for general
independent distribution, and E for Erlang distribution.

15.2 Balance Equations and Little's Law Equations for *B&D* Queueing Models

Definition 4 (*Balance equations for a general B&D queueing model with infinite
capacity*) The state transition diagram in Fig. 15.1 describes a general *B&D* queueing
model with infinite capacity.

The balance equations in Table 15.2 also describe a general *B&D* queueing model
with infinite capacity.

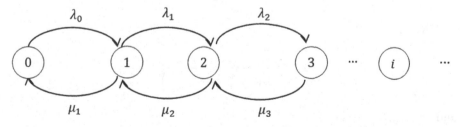

Fig. 15.1 State transition diagram for a general *B&D* queueing model with infinite capacity

Table 15.2 Balance equations for a general *B&D* queueing model with infinite capacity

State	Rate at which B&D process leaves state = Rate at which B&D process enters state
0	$\lambda_0 P_0 = \mu_1 P_1$
1	$(\lambda_1 + \mu_1) P_1 = \lambda_0 P_0 + \mu_2 P_2$
2	$(\lambda_2 + \mu_2) P_2 = \lambda_1 P_1 + \mu_3 P_3$
$i \geq 1$	$(\lambda_i + \mu_i) P_i = \lambda_{i-1} P_{i-1} + \mu_{i+1} P_{i+1}$

Note that from the balance equations

$$P_1 = \left(\frac{\lambda_0}{\mu_1}\right) P_0$$

$$P_2 = \left(\frac{\lambda_1}{\mu_2}\right) P_1 + \left(\frac{\mu_1}{\mu_2}\right) P_1 - \left(\frac{\lambda_0}{\mu_2}\right) P_0 = \left(\frac{\lambda_1}{\mu_2}\right) P_1 = \left(\frac{\lambda_1}{\mu_2}\right)\left(\frac{\lambda_0}{\mu_1}\right) P_0 = \left(\frac{\lambda_1 \lambda_0}{\mu_2 \mu_1}\right) P_0$$

and generally

$$P_i = \left(\frac{\lambda_{i-1} \ldots \lambda_1 \lambda_0}{\mu_i \ldots \mu_2 \mu_1}\right) P_0 \quad \text{for } i \geq 1$$

will be derived. Since

$$\sum_{i=0}^{\infty} P_i = 1$$

it follows that

$$P_0 + \sum_{i=1}^{\infty} P_i = 1$$

and

$$P_0 + \sum_{i=1}^{\infty} \left(\frac{\lambda_{i-1} \ldots \lambda_1 \lambda_0}{\mu_i \ldots \mu_2 \mu_1}\right) P_0 = 1$$

Finally,

$$P_0 = \frac{1}{1 + \sum_{i=1}^{\infty} \left(\frac{\lambda_{i-1} \ldots \lambda_1 \lambda_0}{\mu_i \ldots \mu_2 \mu_1}\right)}$$

and

$$P_i = \frac{\left(\frac{\lambda_{i-1}...\lambda_1 \lambda_0}{\mu_i...\mu_2 \mu_1}\right)}{1 + \sum_{i=1}^{\infty}\left(\frac{\lambda_{i-1}...\lambda_1 \lambda_0}{\mu_i...\mu_2 \mu_1}\right)} \quad \text{for } i \geq 1$$

will hold, where

$$\sum_{i=1}^{\infty}\left(\frac{\lambda_{i-1}...\lambda_1 \lambda_0}{\mu_i...\mu_2 \mu_1}\right) < \infty$$

Example 3 There is a system with 3 servers. The service time at each server is *exponentially* distributed with mean 5 min. Customers arrive to this system according to a *Poisson process* at a rate of 10 per hour. Any potential customer who sees 1 customer waiting in queue cannot enter the system.

According to the Kendall's notation, this is an $M/M/3/4$ queueing model. The state transition diagram in Fig. 15.2 describes the system.

Note that the service rate for any server is $\mu = \frac{60}{5} = 12$ per hour. Based on Definition 7 in Chap. 14 provided for the $M/M/m$ queueing models, $\mu_1, \mu_2, \mu_3, \mu_4$ have been calculated as shown on the state transition diagram in Fig. 15.2.

The balance equations in Table 15.3 also describe the system.

From the first balance equation, it is clear that

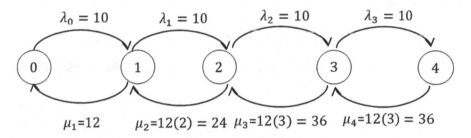

Fig. 15.2 State transition diagram for the $M/M/3/4$ queueing model in example 3

Table 15.3 Balance equations for the $M/M/3/4$ queueing model in example 3

State	Rate at which B&D process leaves state = Rate at which B&D process enters state
0	$10P_0 = 12P_1$
1	$(12 + 10)P_1 = 10P_0 + 24P_2$
2	$(24 + 10)P_2 = 10P_1 + 36P_3$
3	$(36 + 10)P_3 = 10P_2 + 36P_4$
4	$36P_4 = 10P_3$

$$P_1 = \left(\frac{10}{12}\right)P_0 = \left(\frac{5}{6}\right)P_0 \cong 0.8333\,P_0$$

By using the other balance equations, P_2, P_3, P_4 can also be written in terms of P_0 as follows:

$$22(0.8333\,P_0) = 10P_0 + 24P_2$$
$$P_2 \cong 0.3472\,P_0$$
$$34(0.3472\,P_0) = 10(0.8333\,P_0) + 36P_3$$
$$P_3 \cong 0.0965\,P_0$$
$$36P_4 = 10(0.0965\,P_0)$$
$$P_4 \cong 0.0268\,P_0$$

Note that by using

$$P_0 + P_1 + P_2 + P_3 + P_4 = 1$$

and

$$P_0 + 0.8333\,P_0 + 0.3472\,P_0 + 0.0965\,P_0 + 0.0268\,P_0 = 1$$

we can compute

$$P_0 = 0.4341$$

By using the abovementioned equations, P_1, P_2, P_3, P_4 can also be calculated as follows:

$$P_1 = 0.3617$$
$$P_2 = 0.1507$$
$$P_3 = 0.0419$$
$$P_4 = 0.0116$$

Accordingly, the average number of customers in the system will be

$$L = 0P_0 + 1P_1 + 2P_2 + 3P_3 + 4P_4$$
$$= 1(0.3617) + 2(0.1507) + 3(0.0419) + 4(0.0116) = 0.8352$$

The proportion of potential customers that enter the system will be

$$P_0 + P_1 + P_2 + P_3 = 1 - P_4 = 1 - 0.0116 = 0.9884$$

The average amount of time an *entering customer* spends in the system will be

$$W = \frac{L}{\lambda_a}$$

from *Little's law*, where

$$\lambda_a = \lambda(1 - P_4) = 10(0.9884) = 9.884$$

Finally,

$$W = \frac{L}{\lambda_a} = \frac{0.8352}{9.884} \cong 0.084\,\text{h} \cong 5.04\,\text{min}$$

Definition 5 (*Balance equations for an M/M/1 queueing model with infinite capacity*) The state transition diagram in Fig. 15.3 describes an *M/M/1* queueing model with infinite capacity.

The balance equations in Table 15.4 also describe an *M/M/1* queueing model with infinite capacity.

Note that from the balance equations

$$P_1 = \left(\frac{\lambda}{\mu}\right) P_0$$

$$P_2 = \left(\frac{\lambda}{\mu}\right) P_1 + \left(\frac{\mu}{\mu}\right) P_1 - \left(\frac{\lambda}{\mu}\right) P_0 = \left(\frac{\lambda}{\mu}\right) P_1 = \left(\frac{\lambda}{\mu}\right)\left(\frac{\lambda}{\mu}\right) P_0 = \left(\frac{\lambda}{\mu}\right)^2 P_0$$

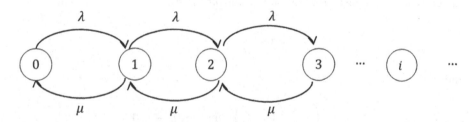

Fig. 15.3 State transition diagram for an *M/M/1* queueing model with infinite capacity

Table 15.4 Balance equations for an *M/M/1* queueing model with infinite capacity

State	Rate at which B&D process leaves state = Rate at which B&D process enters state
0	$\lambda P_0 = \mu P_1$
1	$(\lambda + \mu) P_1 = \lambda P_0 + \mu P_2$
2	$(\lambda + \mu) P_2 = \lambda P_1 + \mu P_3$
$i \geq 1$	$(\lambda + \mu) P_i = \lambda P_{i-1} + \mu P_{i+1}$

and generally

$$P_i = \left(\frac{\lambda}{\mu}\right)^i P_0 \quad \text{for } i \geq 0$$

will be derived. Since

$$\sum_{i=0}^{\infty} P_i = 1$$

holds, it follows that

$$\sum_{i=0}^{\infty} \left(\frac{\lambda}{\mu}\right)^i P_0 = 1$$

and

$$P_0 \sum_{i=0}^{\infty} \left(\frac{\lambda}{\mu}\right)^i = 1$$

Accordingly,

$$P_0 \left(\frac{1}{1 - \frac{\lambda}{\mu}}\right) = 1$$

can be written by considering the sum of the *infinite geometric series*. Finally,

$$P_0 = 1 - \frac{\lambda}{\mu} \quad \text{for } \frac{\lambda}{\mu} < 1$$

and

$$P_i = \left(\frac{\lambda}{\mu}\right)^i \left(1 - \frac{\lambda}{\mu}\right) \quad \text{for } i \geq 0, \frac{\lambda}{\mu} < 1$$

Definition 6 (*Little's law equations for an M/M/1 queueing model with infinite capacity*) Since

$$P_i = \left(\frac{\lambda}{\mu}\right)^i \left(1 - \frac{\lambda}{\mu}\right) \quad \text{for } i \geq 0, \frac{\lambda}{\mu} < 1$$

holds from Definition 5,

L: Average number of objects in the system

can be calculated as

$$L = \sum_{i=0}^{\infty} i\, P_i = \sum_{i=0}^{\infty} i \left(\frac{\lambda}{\mu}\right)^i \left(1 - \frac{\lambda}{\mu}\right) = \frac{\lambda}{\mu - \lambda}$$

W: Average amount of time an object spends in the system

can be calculated as

$$W = \frac{L}{\lambda_a} = \frac{L}{\lambda} = \frac{\left(\frac{\lambda}{\mu-\lambda}\right)}{\lambda} = \frac{1}{\mu - \lambda}$$

W_Q: Average amount of time an object spends waiting in queue

can be calculated as

$$W_Q = W - E[X] = \frac{1}{\mu - \lambda} - \frac{1}{\mu} = \frac{\lambda}{\mu(\mu - \lambda)}$$

where

$$E[X] = \frac{1}{\mu}$$

is the expected service time. Finally,

L_Q: Average number of objects waiting in queue

can be calculated as

$$L_Q = \lambda_a W_Q = \lambda W_Q = \lambda \left(\frac{\lambda}{\mu(\mu - \lambda)}\right) = \frac{\lambda^2}{\mu(\mu - \lambda)}$$

Remark 3 $L = \sum_{i=0}^{\infty} i\left(\frac{\lambda}{\mu}\right)^i \left(1 - \frac{\lambda}{\mu}\right) = \frac{\lambda}{\mu-\lambda}$ follows due to the identity $\sum_{i=0}^{\infty} i a^i = \frac{a}{(1-a)^2}$.

Example 4 People arrive to a bank with 1 server according to a *Poisson process* at a rate of 10 per hour, and the service time is *exponentially* distributed with mean 5 min.

This system is an $M/M/1$ queueing model with parameters

$$\lambda = 10 \, \text{per hour}$$
$$\mu = \frac{60}{5} = 12 \, \text{per hour}$$

Accordingly, the limiting probability that there is *no customer in the system* will be

$$P_0 = 1 - \frac{\lambda}{\mu} = 1 - \frac{10}{12} = \frac{1}{6}$$

the average number of customers in the bank will be

$$L = \frac{\lambda}{\mu - \lambda} = \frac{10}{12 - 10} = 5$$

the average number of customers waiting in queue will be

$$L_Q = \frac{\lambda^2}{\mu(\mu - \lambda)} = \frac{(10)^2}{12(12 - 10)} = \frac{25}{6} \cong 4.17$$

the average amount of time a customer spends in the bank will be

$$W = \frac{1}{\mu - \lambda} = \frac{1}{12 - 10} = \frac{1}{2}h = 30\,\text{min}$$

and the average amount of time a customer spends waiting in queue will be

$$W_Q = \frac{\lambda}{\mu(\mu - \lambda)} = \frac{10}{12(12 - 10)} = \frac{5}{12}h = 25\,\text{min}$$

Definition 7 (*Balance equations for an M/M/1/K queueing model*) The state transition diagram in Fig. 15.4 describes an $M/M/1/K$ queueing model, which is an $M/M/1$ queueing model with K capacity.

The balance equations in Table 15.5 also describe an $M/M/1/K$ queueing model. Note that from the balance equations

$$P_0 = \frac{\left(1 - \frac{\lambda}{\mu}\right)}{1 - \left(\frac{\lambda}{\mu}\right)^{K+1}}$$

and

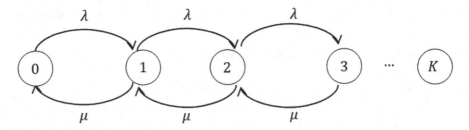

Fig. 15.4 State transition diagram for an *M/M/1/K* queueing model

Table 15.5 Balance equations for an *M/M/1/K* queueing model

State	Rate at which B&D process leaves state = Rate at which B&D process enters state
0	$\lambda P_0 = \mu P_1$
$1 \le i \le K-1$	$(\lambda + \mu) P_i = \lambda P_{i-1} + \mu P_{i+1}$
K	$\mu P_K = \lambda P_{K-1}$

$$P_i = \frac{\left(\frac{\lambda}{\mu}\right)^i \left(1 - \frac{\lambda}{\mu}\right)}{1 - \left(\frac{\lambda}{\mu}\right)^{K+1}} \quad \text{for } i = 0, 1, 2, \ldots, K$$

will follow.

Definition 8 (*Little's law equations for an M/M/1/K queueing model*) It can be shown that

$$L = \frac{\lambda\left[1 + K(\lambda/\mu)^{K+1} - (K+1)(\lambda/\mu)^K\right]}{(\mu - \lambda)\left(1 - (\lambda/\mu)^{K+1}\right)}$$

is the average number of objects in the system. Additionally,

$$W = \frac{L}{\lambda_a}$$

$$W_Q = W - E[X] = W - \frac{1}{\mu}$$

$$L_Q = \lambda_a W_Q$$

are the other equations, where

> X : Service time
>
> $\lambda_a = \lambda$ if we consider all *arriving customers*
>
> $\lambda_a = \lambda(1 - P_K)$ if we consider only *entering customers*

Example 5 We consider an *M/M/1/4* queueing model with mean interarrival times of 12 min, and a service rate of 10 per hour. The parameters of this queueing model will be

$$\lambda = \frac{60}{12} = 5 \text{ per hour}$$

$$\mu = 10 \text{ per hour}$$

$$K = 4$$

Accordingly, the limiting probability that there is *no* customer in the system will be

$$P_0 = \frac{\left(1 - \frac{\lambda}{\mu}\right)}{1 - \left(\frac{\lambda}{\mu}\right)^{K+1}} = \frac{\left(1 - \frac{5}{10}\right)}{1 - \left(\frac{5}{10}\right)^5} = 0.5161$$

the limiting probability that there are 4 customers in the system will be

$$P_4 = \frac{\left(\frac{\lambda}{\mu}\right)^4 \left(1 - \frac{\lambda}{\mu}\right)}{1 - \left(\frac{\lambda}{\mu}\right)^{K+1}} = \frac{\left(\frac{5}{10}\right)^4 \left(1 - \frac{5}{10}\right)}{1 - \left(\frac{5}{10}\right)^5} = 0.0323$$

and the average number of customers in the system will be

$$L = \frac{\lambda \left[1 + K(\lambda/\mu)^{K+1} - (K + 1)(\lambda/\mu)^K\right]}{(\mu - \lambda)\left(1 - (\lambda/\mu)^{K+1}\right)}$$

$$= \frac{5\left[1 + 4(5/10)^5 - 5(5/10)^4\right]}{(10 - 5)\left(1 - (5/10)^5\right)} = 0.8387$$

If we consider all *arriving customers*, then the average amount of time a customer spends in the system will be

$$W = \frac{L}{\lambda_a} = \frac{L}{\lambda} = \frac{0.8387}{5} = 0.1677\,\text{h} = 10.06\,\text{min}$$

If we consider only *entering customers*, then the average amount of time a customer spends in the system will be

$$W = \frac{L}{\lambda_a} = \frac{L}{\lambda(1 - P_4)} = \frac{0.8387}{5(1 - 0.0323)} = 0.1733\,\text{h} = 10.4\,\text{min}$$

where $P_4 = 0.0323$.

If we consider all *arriving customers*, then the average amount of time a customer spends in queue will be

$$W_Q = W - \frac{1}{\mu} = 0.1677 - \frac{1}{10} = 0.0677\,\text{h} = 4.06\,\text{min}$$

If we consider only *entering customers*, then the average amount of time a customer spends in queue will be

$$W_Q = W - \frac{1}{\mu} = 0.1733 - \frac{1}{10} = 0.0733\,\text{h} = 4.4\,\text{min}$$

If we consider all *arriving customers*, then the average number of customers waiting in queue will be

$$L_Q = \lambda_a W_Q = \lambda W_Q = 5(0.0677) = 0.3385$$

If we consider only *entering customers*, then the average number of customers waiting in queue will be

$$L_Q = \lambda_a W_Q = \lambda(1 - P_4)W_Q = 5(1 - 0.0323)0.0733 = 0.3546$$

Problems

1. There is a one-server system with a capacity of 3 customers. Customers arrive to this system according to a *Poisson process* at a rate of 5 per hour. Service times are *exponentially* distributed with mean 10 min.

   (a) Find the average number of customers in the system.
   (b) Find the proportion of potential customers that enter the system.
   (c) Find the average amount of time an entering customer spends in the system.

Solution
This is an $M/M/1/3$ queueing model with parameters

$$\lambda = 5 \text{ per hour}$$
$$\mu = \frac{60}{10} = 6 \text{ per hour}$$
$$K = 3$$

The state transition diagram in Fig. 15.5 describes the system. The balance equations in Table 15.6 also describe the system. Note that from the balance equations, it follows that

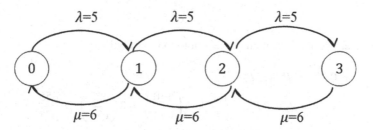

Fig. 15.5 State transition diagram for the $M/M/1/3$ queueing model in problem 1

Table 15.6 Balance equations for the $M/M/1/3$ queueing model in problem 1

State	Rate at which B&D process leaves state = Rate at which B&D process enters state
0	$5P_0 = 6P_1$
1	$11P_1 = 5P_0 + 6P_2$
2	$11P_2 = 5P_1 + 6P_3$
3	$6P_3 = 5P_2$

$$P_1 = \left(\frac{5}{6}\right) P_0$$

$$11\left(\frac{5}{6}P_0\right) = 5P_0 + 6P_2$$

$$P_2 = \left(\frac{25}{36}\right) P_0$$

$$P_3 = \left(\frac{5}{6}\right) P_2 = \left(\frac{125}{216}\right) P_0$$

Since

$$P_0 + P_1 + P_2 + P_3 = 1$$

$$P_0 + \left(\frac{5}{6}\right) P_0 + \left(\frac{25}{36}\right) P_0 + \left(\frac{125}{216}\right) P_0 = 1$$

we can compute the limiting probabilities as

$$P_0 = \frac{216}{671}$$

$$P_1 = \frac{180}{671}$$

$$P_2 = \frac{150}{671}$$

$$P_3 = \frac{125}{671}$$

(a) The average number of customers in the system will be

$$L = 0P_0 + 1P_1 + 2P_2 + 3P_3$$
$$= P_1 + 2P_2 + 3P_3 = \frac{180}{671} + 2\left(\frac{150}{671}\right) + 3\left(\frac{125}{671}\right) = \frac{855}{671} \cong 1.274$$

(b) The proportion of potential customers that enter the system will be

$$P_0 + P_1 + P_2 = 1 - P_3 = 1 - \left(\frac{125}{671}\right) = \frac{546}{671} \cong 0.814$$

(c) The average amount of time an *entering customer* spends in the system will be

$$W = \frac{L}{\lambda_a} = \frac{L}{\lambda(1 - P_3)} = \frac{1.274}{5(0.814)} \cong 0.313\,\text{h} = 18.78\,\text{min}$$

2. Consider a repair system with two repairmen who work for fixing turning lathes in a jobshop with 3 turning lathes. The amount of time each turning lathe functions before breaking down is *exponentially* distributed with mean 250 h. The amount of time it takes each repairman to fix a turning lathe is *exponentially* distributed with mean 6 h.

   (a) Find the average number of the non-functioning turning lathes.
   (b) Find the average amount of time each turning lathe spends in the repair system.
   (c) Find the proportion of time that both repairmen are busy.

Solution
The repair system is a general *B&D* queueing model with the state transition diagram given in Fig. 15.6.
   Note that $\lambda_0 = \frac{1}{250} + \frac{1}{250} + \frac{1}{250} = \frac{3}{250}$ is the arrival rate for the time to the *first failure* when there is no turning lathe in the repair system, i.e. all 3 turning lathes are functioning. The explanations for λ_1 and λ_2 will be analogous. $\mu_2 - \frac{1}{6} + \frac{1}{6} = \frac{2}{6}$ is the departure rate for the time to the *first departure* from the repair system when there are 2 turning lathes in the repair system. $\mu_3 = \frac{2}{6}$ will also follow since there are only 2 turning lathes being fixed, and there is 1 turning lathe waiting in queue. The computations of all these arrival rates and departure rates follow from *further property* 1 of exponential random variable.
   The balance equations in Table 15.7 describe the system.
   By using the balance equations, we can find

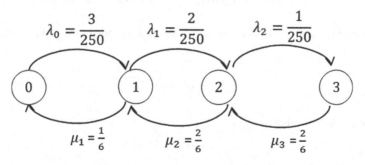

Fig. 15.6 State transition diagram for the repair system in problem 2

Table 15.7 Balance equations for the repair system in problem 2

State	Rate at which B&D process leaves state = Rate at which B&D process enters state
0	$\left(\frac{3}{250}\right)P_0 = \left(\frac{1}{6}\right)P_1$
1	$\left(\frac{2}{250} + \frac{1}{6}\right)P_1 = \left(\frac{3}{250}\right)P_0 + \left(\frac{2}{6}\right)P_2$
2	$\left(\frac{1}{250} + \frac{2}{6}\right)P_2 = \left(\frac{2}{250}\right)P_1 + \left(\frac{2}{6}\right)P_3$
3	$\left(\frac{2}{6}\right)P_3 = \left(\frac{1}{250}\right)P_2$

$$P_1 = 0.072 P_0$$
$$P_2 = 0.0017 P_0$$
$$P_3 = 0.00002 P_0$$

Since

$$P_0 + P_1 + P_2 + P_3 = 1$$
$$P_0 + 0.072 P_0 + 0.0017 P_0 + 0.00002 P_0 = 1$$

we can compute

$$P_0 = 0.9313$$
$$P_1 = 0.0671$$
$$P_2 = 0.0016$$
$$P_3 = 0.00002$$

(a) The average number of the non-functioning turning lathes will be the average number of the turning lathes in the repair system such that

$$L = 0P_0 + 1P_1 + 2P_2 + 3P_3$$
$$= P_1 + 2P_2 + 3P_3$$
$$= 0.0671 + 2(0.0016) + 3(0.00002) = 0.0704$$

(b) The average amount of time each turning lathe spends in the repair system will be

$$W = \frac{L}{\lambda_a}$$

by *Little's law*, where

$$\lambda_a = \lambda_0 P_0 + \lambda_1 P_1 + \lambda_2 P_2$$

Table 15.8 States for the $M/M/1$ system with *bulk* service in problem 3

State	Explanation
$0'$	No one is in service
0	Server is busy, no one is waiting to be served
$i \geq 1$	i customers are waiting to be served

$$= \left(\frac{3}{250}\right)(0.9313) + \left(\frac{2}{250}\right)(0.0671) + \left(\frac{1}{250}\right)(0.0016) = 0.0117$$

is the *state-dependent arrival rate*.
Finally,

$$W = \frac{L}{\lambda_a} = \frac{0.0704}{0.0117} \cong 6.0171\,\text{h}$$

(c) The proportion of time that both repairmen are busy will be

$$P_2 + P_3 = 0.0016 + 0.00002 = 0.00162$$

3. Consider an $M/M/1$ system with *bulk* service, in which the server can serve at most 3 customers simultaneously. The arrival rate of customers is λ, and the service rate is μ. Build the state transition diagram, and the balance equations for this queueing model.[1]

Solution
We consider the states given in Table 15.8 for the $M/M/1$ system with *bulk* service.
Accordingly, we can build the state transition diagram in Fig. 15.7.
We can also build the balance equations provided in Table 15.9.

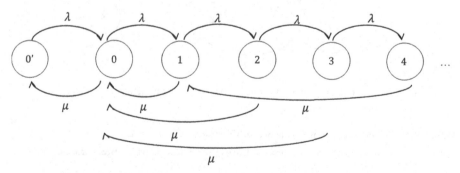

Fig. 15.7 State transition diagram for the $M/M/1$ system with *bulk* service in problem 3

---

[1]This problem and its solution approach have been borrowed from Introduction to Probability Models by Sheldon Ross.

Table 15.9 Balance equations for the $M/M/1$ system with *bulk* service in problem 3

State	Rate at which B&D process leaves state = Rate at which B&D process enters state
$0'$	$\lambda P_{0'} = \mu P_0$
0	$(\lambda + \mu) P_0 = \lambda P_{0'} + \mu P_1 + \mu P_2 + \mu P_3$
$i \geq 1$	$(\lambda + \mu) P_i = \lambda P_{i-1} + \mu P_{i+3}$

4. Consider a one-server system with an *exponential* service rate μ. Customers arrive to this system according to a *Poisson process* with rate λ, and enter the system with probability $(0.7)^i$ whenever there are i customers. Build the state transition diagram, and the balance equations for this system. Find the limiting probabilities that there is *no* customer in the system, and there are i customers in the system.

Solution

The transition diagram in Fig. 15.8 describes the system:

The balance equations in Table 15.10 also describe the system.

Note that from the balance equations,

$$P_1 = \left(\frac{\lambda_0}{\mu}\right) P_0 = (0.7)^0 \left(\frac{\lambda}{\mu}\right) P_0$$

$$P_2 = \left(\frac{\lambda_1}{\mu}\right) P_1 + \left(\frac{\mu}{\mu}\right) P_1 - \left(\frac{\lambda_0}{\mu}\right) P_0 = \left(\frac{\lambda_1}{\mu}\right) P_1 = (0.7)^0 (0.7)^1 \left(\frac{\lambda}{\mu}\right)^2 P_0$$

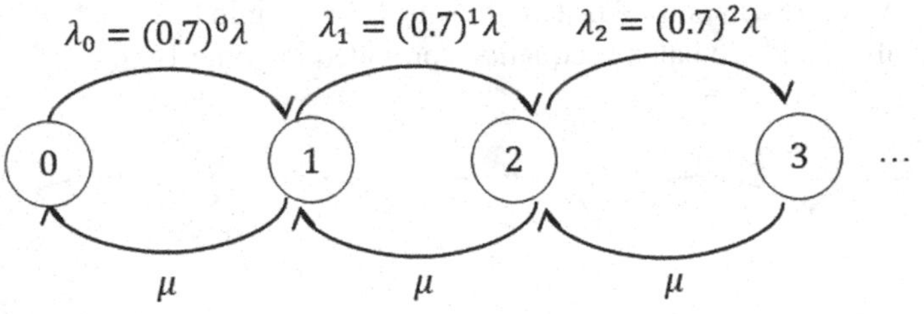

Fig. 15.8 State transition diagram for the one-server system in problem 4

Table 15.10 Balance equations for the one-server system in problem 4

State	Rate at which B&D process leaves state = Rate at which B&D process enters state
0	$\lambda_0 P_0 = \mu P_1$
1	$(\lambda_1 + \mu) P_1 = \lambda_0 P_0 + \mu P_2$
$i \geq 1$	$(\lambda_i + \mu) P_i = \lambda_{i-1} P_{i-1} + \mu P_{i+1}$

and generally

$$P_i = (0.7)^0 (0.7)^1 \ldots (0.7)^{i-1} \left(\frac{\lambda}{\mu}\right)^i P_0 = (0.7)^{\frac{(i-1)i}{2}} \left(\frac{\lambda}{\mu}\right)^i P_0$$

will be derived for $i \geq 1$.

Since

$$\sum_{i=0}^{\infty} P_i = 1$$

$$P_0 + \sum_{i=1}^{\infty} P_i = 1$$

$$P_0 + \sum_{i=1}^{\infty} (0.7)^{\frac{(i-1)i}{2}} \left(\frac{\lambda}{\mu}\right)^i P_0 = 1$$

the limiting probability that there is *no* customer in the system will be

$$P_0 = \frac{1}{1 + \sum_{i=1}^{\infty} (0.7)^{\frac{(i-1)i}{2}} \left(\frac{\lambda}{\mu}\right)^i}$$

and the limiting probability that there are i customers in the system will be

$$P_i = \frac{(0.7)^{\frac{(i-1)i}{2}} \left(\frac{\lambda}{\mu}\right)^i}{1 + \sum_{i=1}^{\infty} (0.7)^{\frac{(i-1)i}{2}} \left(\frac{\lambda}{\mu}\right)^i}$$

5. Show that a *B&D* queueing model is a time-reversible Markov chain.

Solution

Recall the time-reversibility condition for a *CTMC* as

$$P_i q_{ij} = P_j q_{ji} \text{ for all } i, j \in S$$

from Definition 11 in Chap. 14. Accordingly,

$$P_i q_{i,i+1} = P_{i+1} q_{i+1,i} \text{ for } i \geq 0$$
$$P_i q_{i,i-1} = P_{i-1} q_{i-1,i} \text{ for } i > 0$$

should hold for the time-reversibility of a *B&D* queueing model. We will verify that $P_i q_{i,i+1} = P_{i+1} q_{i+1,i}$ holds for $i \geq 0$. The verification of $P_i q_{i,i-1} = P_{i-1} q_{i-1,i}$ for $i > 0$ is left to the reader.

By recalling that

$$P_i = \left(\frac{\lambda_{i-1} \ldots \lambda_1 \lambda_0}{\mu_i \ldots \mu_2 \mu_1} \right) P_0 \quad \text{for } i \geq 1$$

is the limiting probability of state i for a *B&D* queueing model from Definition 4 in this chapter, and by considering the formula for the instantaneous transition rate

$$q_{ik} = v_i P_{ik}$$

from Table 14.7 in Chap. 14, it follows that

$$
\begin{aligned}
P_i q_{i,i+1} &= \left(\frac{\lambda_{i-1} \ldots \lambda_1 \lambda_0}{\mu_i \ldots \mu_2 \mu_1} \right) P_0 v_i P_{i,i+1} \\
&= \left(\frac{\lambda_{i-1} \ldots \lambda_1 \lambda_0}{\mu_i \ldots \mu_2 \mu_1} \right) P_0 (\lambda_i + \mu_i) \frac{\lambda_i}{\lambda_i + \mu_i} \\
&= \left(\frac{\lambda_{i-1} \ldots \lambda_1 \lambda_0}{\mu_i \ldots \mu_2 \mu_1} \right) P_0 \lambda_i = \left(\frac{\lambda_i \ldots \lambda_1 \lambda_0}{\mu_i \ldots \mu_2 \mu_1} \right) P_0
\end{aligned}
$$

Analogously,

$$
\begin{aligned}
P_{i+1} q_{i+1,i} &= \left(\frac{\lambda_i \ldots \lambda_1 \lambda_0}{\mu_{i+1} \ldots \mu_2 \mu_1} \right) P_0 v_{i+1} P_{i+1,i} \\
&= \left(\frac{\lambda_i \ldots \lambda_1 \lambda_0}{\mu_{i+1} \ldots \mu_2 \mu_1} \right) P_0 (\lambda_{i+1} + \mu_{i+1}) \frac{\mu_{i+1}}{\lambda_{i+1} + \mu_{i+1}} \\
&= \left(\frac{\lambda_i \ldots \lambda_1 \lambda_0}{\mu_{i+1} \ldots \mu_2 \mu_1} \right) P_0 \mu_{i+1} = \left(\frac{\lambda_i \ldots \lambda_1 \lambda_0}{\mu_i \ldots \mu_2 \mu_1} \right) P_0
\end{aligned}
$$

As a result,

$$P_i q_{i,i+1} = P_{i+1} q_{i+1,i} \quad \text{for } i \geq 0$$

Chapter 16
Introduction to Brownian Motion

Abstract Brownian motion (BM) as a continuous-time extension to a simple symmetric random walk has been introduced in this chapter. The stationary and independent increments, normal distribution, and Markovian property have been provided as the properties of a standard Brownian motion. Brownian motion with drift, and geometric Brownian motion have also been defined as an extension to a standard Brownian motion. Some examples and problems have been provided to clarify the basic and other properties of a Brownian motion.

Brownian motion (BM) is a continuous-time extension of a simple symmetric random walk introduced in Chap. 13. Analogous to a homogeneous Poisson process introduced in Chaps. 9 and 10, BM possesses stationary and independent increments. Additionally, each random variable of a BM is a normal random variable which was defined in Chap. 6, and BM possesses Markovian property which was defined in Chap. 12. As a special case of a BM, each random variable of a geometric Brownian motion is a lognormal random variable which was introduced in Chap. 6. Briefly, the contents of this chapter are closely related to several concepts from the other chapters. Particularly, the importance of the normal random variable and standard normal random variable from Chap. 6 will be clear in the examples and problems of this chapter.

16.1 Basic Properties of a Brownian Motion

Definition 1 (*Brownian Motion—BM*) A simple symmetric random walk, an extension to a simple symmetric random walk, and Brownian motion are defined in comparison to each other in Table 16.1.

Remark 1 Recall that Δ_i' s are *i.i.d.*, and

$$E[\Delta_i] = 0 \quad Var(\Delta_i) = 1 \text{ for all } i$$

for a simple symmetric random walk from Table 13.1 in Chap. 13. Thus,

© Springer Nature Switzerland AG 2019

E. Bas, *Basics of Probability and Stochastic Processes*,

https://doi.org/10.1007/978-3-030-32323-3_16

Table 16.1 Simple symmetric random walk, extension to a simple symmetric random walk, and Brownian motion

Simple symmetric random walk	Extension to a simple symmetric random walk
Let $$X_n = \sum_{i=1}^{n} \Delta_i$$ be defined, where for all $i \geq 1$ $\Delta_i = 1$ with probability $1/2$ or $\Delta_i = -1$ with probability $1/2$ Hence, $\{X_n; n \geq 0\}$ is a *simple symmetric random walk*.	Let $$B_k(t) = \frac{1}{\sqrt{k}} \sum_{i=1}^{tk} \Delta_i, \ k > 1$$ be defined, where for all $i \geq 1$ $\Delta_i = 1$ with probability $1/2$ or $\Delta_i = -1$ with probability $1/2$ Hence, $\{B_k(t); t \geq 0\}$ is a *CTMC* as an *extension to a simple symmetric random walk*.
X_n can be interpreted as the position of a particle at minute n. In a simple symmetric random walk, the *step size* is 1, and the movements are assumed to be made *at every* 1 min. Note that $E[X_n] = 0$ $Var(X_n) = n$	$B_k(t)$ can be interpreted as the position of a particle at minute t. In an extension to a simple symmetric random walk, the *step size* is $1/\sqrt{k}$, and the movements are assumed to be made at *every* $1/k$ minute for a *very large integer number* k. Note that $E[B_k(t)] = 0$ $Var(B_k(t)) = t$
	Brownian motion
	If $k \to \infty$, then $\{B_k(t); t \geq 0\}$ will converge to $\{B(t); t \geq 0\}$ which is called as *Brownian motion (BM)*, where $E[B(t)] = 0$ $Var(B(t)) = t$

$$E[B_k(t)] = E\left(\frac{1}{\sqrt{k}} \sum_{i=1}^{tk} \Delta_i\right) = \frac{1}{\sqrt{k}} \sum_{i=1}^{tk} E[\Delta_i] = 0$$

$$Var(B_k(t)) = Var\left(\frac{1}{\sqrt{k}} \sum_{i=1}^{tk} \Delta_i\right) = \left(\frac{1}{\sqrt{k}}\right)^2 \sum_{i=1}^{tk} Var(\Delta_i) = \left(\frac{1}{\sqrt{k}}\right)^2 tk = t$$

Definition 2 (*Stationary and independent increments for a Brownian motion*) The definitions of the stationary and independent increments for the extension to a simple symmetric random walk, and for Brownian motion are provided in Table 16.2.

Definition 3 (*B(t) of a Brownian motion as a normal random variable*) It can be shown that

$$B_k(t) \to N(0, t) \ \text{as} \ k \to \infty$$

Table 16.2 Stationary and independent increments for the extension to a simple symmetric random walk, and for Brownian motion

Stationary increments for the extension to a simple symmetric random walk	Stationary increments for Brownian motion
$B_k(t) - B_k(s)$ distribution depends *only* on the *interval length* $t - s$, which means the extension to a simple symmetric random walk possesses *stationary increments*.	$B(t) - B(s)$ distribution depends *only* on the *interval length* $t - s$, which means Brownian motion possesses *stationary increments*.
Independent increments for the extension to a simple symmetric random walk	*Independent increments for Brownian motion*
For any two *nonoverlapping intervals* $(t_1, t_2]$ and $(t_3, t_4]$, the distributions of $B_k(t_2) - B_k(t_1)$ and $B_k(t_4) - B_k(t_3)$ are *independent*; which means the extension to a simple symmetric random walk possesses *independent increments*.	For any two *nonoverlapping intervals* $(t_1, t_2]$ and $(t_3, t_4]$, the distributions of $B(t_2) - B(t_1)$ and $B(t_4) - B(t_3)$ are *independent*; which means Brownian motion possesses *independent increments*.

which means

$$B(t) \sim N(0, t)$$

In other words, for Brownian motion $\{B(t); t \geq 0\}$, $B(t)$ is a *normal random variable* with mean 0 and variance t. Since Brownian motion has *stationary increments* as defined in Table 16.2, $B(t) - B(s)$ is also a *normal random variable* with mean 0 and variance $t - s$.

Remark 2 Recall that

$$B_k(t) = \frac{1}{\sqrt{k}} \sum_{i=1}^{tk} \Delta_i, \ k > 1$$

$$E[B_k(t)] = 0$$

$$Var(B_k(t)) = t$$

from Table 16.1. By considering *Central Limit Theorem*, it follows that

$$\frac{B_k(t) - E[B_k(t)]}{\sqrt{Var(B_k(t))}} = \frac{\left(\frac{1}{\sqrt{k}} \sum_{i=1}^{tk} \Delta_i\right) - 0}{\sqrt{t}}$$

$$= \frac{1}{\sqrt{t}} \left(\frac{1}{\sqrt{k}} \sum_{i=1}^{tk} \Delta_i\right) \rightarrow N(0, 1) \text{ as } k \rightarrow \infty$$

and

$$\frac{1}{\sqrt{t}} B_k(t) \to N(0, 1) \quad \text{as } k \to \infty$$

$$B_k(t) \to \sqrt{t} N(0, 1) \quad \text{as } k \to \infty$$

$$B_k(t) \to N(0, t) \quad \text{as } k \to \infty$$

which means

$$B(t) \sim N(0, t)$$

Example 1 A particle moves according to a Brownian motion $\{B(t); t \geq 0\}$. As an example, the expected value of the position of the particle at time $t = 10$ will be

$$E[B(10)] = 0$$

and the variance of the position of the particle at time $t = 10$ will be

$$Var(B(10)) = 10$$

which means

$$B(10) \sim N(0, 10)$$

Given that the particle's position is 5 at time $t = 6$, its expected position at time $t = 9$ will be

$$
\begin{aligned}
E(B(9)|B(6) = 5) &= E(B(6) + B(9) - B(6)|B(6) = 5) \\
&= E(B(6)|B(6) = 5) + E(B(9) - B(6)|B(6) = 5) \\
&= 5 + E(B(9) - B(6)) = 5 + E[B(3)] = 5 + 0 = 5
\end{aligned}
$$

Remark 3 Note that $E(B(9) - B(6)|B(6) = 5) = E(B(9) - B(6))$ follows due to the *independent increments* of two nonoverlapping intervals $(0, 6]$ and $(6, 9]$, and $E(B(9) - B(6)) = E[B(3)]$ due to the *stationary increments* of a Brownian motion.

16.2 Other Properties of a Brownian Motion

Definition 4 (*Brownian motion with Markovian property*) Due to the *independent increments* of a Brownian motion, the future $B(s + t)$, given the present state $B(s)$ only depends on $B(s + t) - B(s)$, which is *independent of the past*. This property means that a Brownian motion has *Markovian* property.

Definition 5 (*Brownian motion with drift*)

$$\{X(t); t \geq 0\}$$

is called as a Brownian motion with variance σ^2 and drift μ, where

$$X(t) = \sigma B(t) + \mu t$$

Since $B(t)$ is a *normal random variable* with mean 0 and variance t, $X(t)$ is also a *normal random variable* with mean μt and variance $\sigma^2 t$. Additionally, since $B(t) - B(s)$ is a *normal random variable* with mean 0 and variance $t - s$, $X(t) - X(s)$ is also a *normal random variable* with mean $\mu(t - s)$ and variance $\sigma^2(t - s)$.

Remark 4 Note that for $\sigma^2 = 1$ and $\mu = 0$,

$$X(t) = B(t)$$

holds, and

$$\{X(t); t \geq 0\}$$

reduces to

$$\{B(t); t \geq 0\}$$

which is called as *standard Brownian motion*.

Example 2 Consider a Brownian motion $\{X(t); t \geq 0\}$ with variance $\sigma^2 = 0.4$ and drift $\mu = 0.8$. The probability that $X(t)$ takes a value between 2 and 5 at time $t = 8$ will be

$$P(2 < X(8) < 5)$$

Recall that

$$X(t) \sim N\left(\mu t, \sigma^2 t\right)$$

Hence,

$$X(8) \sim N((0.8)(8), (0.4)(8)) \text{ and } X(8) \sim N(6.4, 3.2)$$

and

$$P(2 < X(8) < 5) = P\left(\frac{2 - 6.4}{\sqrt{3.2}} < \frac{X(8) - 6.4}{\sqrt{3.2}} < \frac{5 - 6.4}{\sqrt{3.2}}\right)$$

$$= P(-2.4597 < Z < -0.7826) = \Phi(-0.7826) - \Phi(-2.4597)$$
$$= \Phi(2.4597) - \Phi(0.7826) \cong 0.9931 - 0.7823 = 0.2108$$

Definition 6 (*Geometric Brownian motion*)

$$\{S(t); t \geq 0\}$$

is called as a *geometric Brownian motion*, where

$$S(t) = S_0 e^{X(t)}$$

Note that $S_0 > 0$ is the initial value, and $\{X(t); t \geq 0\}$ is a Brownian motion with drift such that

$$X(t) = \sigma B(t) + \mu t$$

Remark 5 $S(t) > 0$ always holds, which makes geometric Brownian motion applicable to model some random phenomena such as *stock prices*.

Remark 6 It can be shown that $S(t)$ is a *lognormal random variable* with expected value

$$E[S(t)] = S_0 e^{\left(\mu + \frac{\sigma^2}{2}\right)t}$$

and variance

$$Var(S(t)) = S_0^2 e^{(2\mu t + \sigma^2 t)} \left(e^{\sigma^2 t} - 1\right)$$

Example 3 A geometric Brownian motion $\{S(t); t \geq 0\}$ is assumed to model the evolution of the stock prices of a company such that

$$S(t) = S_0 e^{B(t)}$$

holds, where $S_0 = \$5$.

By recalling that $\sigma^2 = 1$ and $\mu = 0$ when $X(t) = B(t)$, the expected stock price at time $t = 1$ will be

$$E[S(1)] = S_0 e^{\left(\mu + \frac{\sigma^2}{2}\right)1} = 5e^{(0 + \frac{1}{2})} = 5e^{0.5} = \$8.2436$$

and the variance of the stock price at time $t = 1$ will be

$$Var(S(1)) = S_0^2 e^{(2\mu + \sigma^2)} \left(e^{\sigma^2} - 1\right)$$
$$= (5)^2 \left(e^{(2)(0)+1}\right)\left(e^1 - 1\right)$$

$$= 25e(e - 1) = 116.769$$

Problems

1. Let

$$\{X(t); t \geq 0\}$$

be a Brownian motion with variance $\sigma^2 = 10$ and drift $\mu = 0$.

(a) $P(X(4) \leq 9) =?$
(b) $P(\{X(6) \geq 8\}|\{X(3) = 5\}) =?$
(c) $Var(4X(3) - 5X(5)) =?$
(d) $P(3X(7) - 2X(10) \leq 3) =?$

Solution
Note that

$$X(t) = \sigma B(t) + \mu t$$
$$= \sqrt{10} B(t)$$

(a) Note that $X(4) = \sqrt{10}B(4)$, and

$$P(X(4) \leq 9) = P\left(\sqrt{10}B(4) \leq 9\right) = P\left(B(4) \leq \frac{9}{\sqrt{10}}\right)$$

Note also that

$$B(t) \sim N(0, t)$$

Thus,

$$B(4) \sim N(0, 4)$$

and

$$P\left(B(4) \leq \frac{9}{\sqrt{10}}\right) = P\left(\frac{B(4) - 0}{\sqrt{4}} \leq \frac{\frac{9}{\sqrt{10}} - 0}{\sqrt{4}}\right)$$
$$= P\left(Z \leq \frac{9}{\sqrt{40}}\right) = P(Z \leq 1.4230)$$
$$= \Phi(1.4230) \cong 0.9222$$

(b) Note that $X(6) = \sqrt{10}B(6)$ and $X(3) = \sqrt{10}B(3)$, and

$$
\begin{aligned}
P(\{X(6) \geq 8\}|\{X(3) = 5\}) &= P\left(\left\{\sqrt{10}B(6) \geq 8\right\}\middle|\left\{\sqrt{10}B(3) = 5\right\}\right) \\
&= P\left(\left\{B(6) \geq \frac{8}{\sqrt{10}}\right\}\middle|\left\{B(3) = \frac{5}{\sqrt{10}}\right\}\right) \\
&= P\left(\left\{B(6) - B(3) \geq \frac{8}{\sqrt{10}} - \frac{5}{\sqrt{10}}\right\}\middle|\left\{B(3) = \frac{5}{\sqrt{10}}\right\}\right) \\
&= P\left(B(6) - B(3) \geq \frac{8}{\sqrt{10}} - \frac{5}{\sqrt{10}}\right) \\
&= P\left(B(3) \geq \frac{3}{\sqrt{10}}\right)
\end{aligned}
$$

Note also that the last two equations follow due to the *independent* and the *stationary increments* of a Brownian motion.

Since

$$
B(3) \sim N(0, 3)
$$

it follows that

$$
P\left(B(3) \geq \frac{3}{\sqrt{10}}\right) = P\left(\frac{B(3) - 0}{\sqrt{3}} \geq \frac{\frac{3}{\sqrt{10}} - 0}{\sqrt{3}}\right) = P\left(Z \geq \frac{3}{\sqrt{30}}\right) = P(Z \geq 0.5477)
$$
$$
= 1 - \Phi(0.5477) \cong 1 - 0.7088 = 0.2912
$$

(c) Note that $X(3) = \sqrt{10}B(3)$ and $X(5) = \sqrt{10}B(5)$, and

$$
\begin{aligned}
Var(4X(3) - 5X(5)) &= Var\left(4\sqrt{10}B(3) - 5\sqrt{10}B(5)\right) \\
&= Var\left(-\sqrt{10}B(3) + 5\sqrt{10}B(3) - 5\sqrt{10}B(5)\right) \\
&= Var\left(-\sqrt{10}B(3) - 5\sqrt{10}(B(5) - B(3))\right) \\
&= Var\left(-\sqrt{10}B(3)\right) + Var\left(-5\sqrt{10}(B(5) - B(3))\right) \\
&= \left(-\sqrt{10}\right)^2 Var(B(3)) + \left(-5\sqrt{10}\right)^2 Var(B(5) - B(3)) \\
&= \left(-\sqrt{10}\right)^2 Var(B(3)) + \left(-5\sqrt{10}\right)^2 Var(B(2)) \\
&= 10 Var(B(3)) + 250 Var(B(2)) \\
&= (10)(3) + (250)(2) = 530
\end{aligned}
$$

Note again that $B(3) \sim N(0, 3)$ and $B(2) \sim N(0, 2)$ hold, and the properties of *independent* and *stationary increments* are also applied in the solution of this sub-problem.

(d)   Note that $X(7) = \sqrt{10}B(7)$ and $X(10) = \sqrt{10}B(10)$, and

$$P(3X(7) - 2X(10) \leq 3) = P\left(3\sqrt{10}B(7) - 2\sqrt{10}B(10) \leq 3\right)$$

$$= P\left(\frac{3\sqrt{10}B(7) - 2\sqrt{10}B(10) - 0}{\sqrt{Var\left(3\sqrt{10}B(7) - 2\sqrt{10}B(10)\right)}} \leq \frac{3 - 0}{\sqrt{Var\left(3\sqrt{10}B(7) - 2\sqrt{10}B(10)\right)}}\right)$$

$$= P\left(Z \leq \frac{3}{\sqrt{Var\left(3\sqrt{10}B(7) - 2\sqrt{10}B(10)\right)}}\right)$$

where

$$Var\left(3\sqrt{10}B(7) - 2\sqrt{10}B(10)\right) = Var\left(\sqrt{10}B(7) + 2\sqrt{10}B(7) - 2\sqrt{10}B(10)\right)$$

$$= Var\left(\sqrt{10}B(7) - 2\sqrt{10}(B(10) - B(7))\right)$$

$$= Var\left(\sqrt{10}B(7)\right) + Var\left(-2\sqrt{10}(B(10) - B(7))\right)$$

$$= \left(\sqrt{10}\right)^2 Var(B(7)) + \left(-2\sqrt{10}\right)^2 Var(B(10) - B(7))$$

$$= \left(\sqrt{10}\right)^2 Var(B(7)) + \left(-2\sqrt{10}\right)^2 Var(B(3))$$

$$= 10 Var(B(7)) + 40 Var(B(3))$$

$$= (10)(7) + (40)(3) = 190$$

Finally,

$$P\left(Z \leq \frac{3}{\sqrt{190}}\right) = P(Z \leq 0.2176) = \Phi(0.2176) \cong 0.5871$$

2.   A particle's position is modelled with a standard Brownian motion $\{B(t); t \geq 0\}$.
     If the position of the particle is 12 at $t = 5$, what is the probability that it will be
     greater than 16 at $t = 8$?

Solution

$$P(\{B(8) > 16\}|\{B(5) = 12\}) = \frac{P(\{B(8) > 16\} \cap \{B(5) = 12\})}{P(B(5) = 12)}$$

$$= \frac{P(\{B(8 - 5) > 16 - 12\} \cap \{B(5) = 12\})}{P(B(5) = 12)}$$

$$= \frac{P(\{B(3) > 4\} \cap \{B(5) = 12\})}{P(B(5) = 12)}$$

$$= \frac{P(B(3) > 4)P(B(5) = 12)}{P(B(5) = 12)}$$

$$= P(B(3) > 4)$$

Note that $P(\{B(8) > 16\} \cap \{B(5) = 12\}) = P(\{B(8 - 5) > 16 - 12\} \cap \{B(5) = 12\})$ follows due to the *stationary increments*, and $P(\{B(3) > 4\} \cap \{B(5) = 12\}) = P(B(3) > 4)P(B(5) = 12)$ follows due to the *independent increments* of a Brownian motion.

Note that

$$B(3) \sim N(0, 3)$$

Thus,

$$P(B(3) > 4) = P\left(\frac{B(3) - 0}{\sqrt{3}} > \frac{4 - 0}{\sqrt{3}}\right)$$

$$= P(Z > 2.3094) = 1 - \Phi(2.3094) \cong 1 - 0.9896 = 0.0104$$

3. The stock prices of a company are assumed to follow a geometric Brownian motion $\{S(t); t \geq 0\}$ with $\mu = 1$ and $\sigma^2 = 3$. Assume the initial stock price as \$2. What is the probability that the stock price will be greater than \$20 at $t = 4$?

Solution
Note that

$$S(t) = S_0 e^{X(t)}$$

where $S_0 = \$2$ is the initial value, and

$$X(t) = \sigma B(t) + \mu t = \sqrt{3} B(t) + t$$

Accordingly,

$$S(t) = 2e^{\sqrt{3}B(t) + t}$$

and

$$S(4) = 2e^{\sqrt{3}B(4) + 4}$$

Hence,

$$P(S(4) > 20) = P\left(2e^{\sqrt{3}B(4) + 4} > 20\right) = P\left(e^{\sqrt{3}B(4) + 4} > 10\right)$$

$$= P\left(\sqrt{3}B(4) + 4 > \ln(10)\right) = P\left(\sqrt{3}B(4) + 4 > 2.3026\right)$$

$$= P\left(\sqrt{3}B(4) > -1.6974\right) = P(B(4) > -0.98)$$

By noting that

$$B(4) \sim N(0, 4)$$

$$P\left(\frac{B(4) - 0}{\sqrt{4}} > \frac{-0.98 - 0}{\sqrt{4}}\right) = P(Z > -0.49) = P(Z \leq 0.49) = \Phi(0.49) = 0.6879$$

Chapter 17
Basics of Martingales

Abstract This chapter was a brief introduction to the martingales. In the first part of the chapter; the definitions of a martingale, submartingale, supermartingale, and Doob type martingale were provided, and some examples were given to show how to verify whether a stochastic process is a martingale, submartingale or supermartingale. In the second part of the chapter, Azuma-Hoeffding inequality, Kolmogorov's inequality for submartingales with an extension to martingales, and the martingale convergence theorem have been defined as the selected theorems of the martingales. Two problems were presented to show that a standard Brownian motion and a special type of Poisson process are martingales.

Although martingales can be thought as a standalone topic with very important implications in different areas such as financial engineering, it uses the basic properties of one of the basic topics of probability, i.e. conditional expectation. This chapter is a very brief introduction to the martingales that includes the basic definitions of a martingale, submartingale, supermartingale, and Doob type martingale with some examples, and some selected theorems including Azuma-Hoeffding inequality, Kolmogorov's inequality for submartingales and the extension of Kolmogorov's inequality to martingales, and martingale convergence theorem. Although the basic properties and some additional properties of a Brownian motion have been provided in Chap. 16, it will be shown that a Brownian motion is also a martingale in the problems section of this chapter. It will also be proved that a special type of Poisson process is also a martingale.

17.1 Martingale, Submartingale, Supermartingale, Doob Type Martingale

Definition 1 (*Martingale*) A stochastic process $\{Z_n; n \geq 1\}$ is said to be a *martingale* if

$$E(Z_{n+1}|Z_1, Z_2, \ldots, Z_n) = Z_n \quad \text{for all } n$$

© Springer Nature Switzerland AG 2019
E. Bas, *Basics of Probability and Stochastic Processes*,
https://doi.org/10.1007/978-3-030-32323-3_17

and $E[|Z_n|] < \infty$ for all n

Remark 1 A martingale represents a *fair game*, where Z_n represents the *total fortune* after the nth gamble. Since

$$E(Z_{n+1}|Z_1, Z_2, \ldots, Z_n) = Z_n \text{ for all } n$$

for a martingale $\{Z_n; n \geq 1\}$, by taking the expected value of both sides

$$E\big[E(Z_{n+1}|Z_1, Z_2, \ldots, Z_n)\big] = E[Z_n] \text{ for all } n$$

is obtained which is equivalent to

$$E\big[Z_{n+1}\big] = E[Z_n] \text{ for all } n$$

Note that

$$E\big[E(Z_{n+1}|Z_1, Z_2, \ldots, Z_n)\big] = E\big[Z_{n+1}\big] \text{ for all } n$$

holds due to the identity

$$E[E(X|Y)] = E[X]$$

provided in Table 7.6 in Chap. 7.

In the same manner, the following equations can be verified for all n:

$$E[Z_n] = E\big[Z_{n-1}\big]$$
$$E\big[Z_{n-1}\big] = E\big[Z_{n-2}\big]$$
$$\cdots$$
$$E[Z_2] = E[Z_1]$$

Finally, by considering all the equations, the following equation will be true:

$$E[Z_n] = E[Z_1] \text{ for all } n$$

Example 1 We consider a *simple symmetric random walk* $\{X_n; n \geq 1\}$, where

$$X_n = X_0 + \Delta_1 + \Delta_2 + \cdots + \Delta_i + \cdots + \Delta_n$$

$$X_{n+1} = X_0 + \Delta_1 + \Delta_2 + \cdots + \Delta_i + \cdots + \Delta_n + \Delta_{n+1} = X_n + \Delta_{n+1}$$

By recalling that $E[\Delta_i] = 0$ holds for all i for a simple symmetric random walk, it can be shown that a simple symmetric random walk is a *martingale* since

Table 17.1 Submartingale versus supermartingale

Submartingale	Supermartingale
A stochastic process $\{Z_n; n \geq 1\}$ with $E[|Z_n|] < \infty$ for all n is said to be a *submartingale* if $E(Z_{n+1}|Z_1, Z_2, \ldots, Z_n) \geq Z_n$ for all n	A stochastic process $\{Z_n; n \geq 1\}$ with $E[|Z_n|] < \infty$ for all n is said to be a *supermartingale* if $E(Z_{n+1}|Z_1, Z_2, \ldots, Z_n) \leq Z_n$ for all n

$$E(X_{n+1}|X_1, X_2, \ldots, X_n) = E(X_n + \Delta_{n+1}|X_1, X_2, \ldots, X_n)$$
$$= E(X_n|X_1, X_2, \ldots, X_n) + E(\Delta_{n+1}|X_1, X_2, \ldots, X_n)$$
$$= X_n + E[\Delta_{n+1}] = X_n + 0 = X_n$$

Definition 2 (*Submartingale vs. supermartingale*) A submartingale and super-martingale are defined in comparison to each other as given in Table 17.1.

Remark 2 Note that a *submartingale* represents a *superfair* game, and a *supermartingale* represents a *subfair* game. Note also that

$$E[Z_n] \geq E[Z_1] \text{ for all } n$$

holds for a *submartingale*, and

$$E[Z_n] \leq E[Z_1] \text{ for all } n$$

holds for a *supermartingale*.

Example 2 We consider a *simple random walk* $\{X_n; n \geq 1\}$, where

$$X_n = X_0 + \Delta_1 + \Delta_2 + \cdots + \Delta_i + \cdots + \Delta_n$$

$$X_{n+1} = X_0 + \Delta_1 + \Delta_2 + \cdots + \Delta_i + \cdots + \Delta_n + \Delta_{n+1} = X_n + \Delta_{n+1}$$

We consider also $\Delta_i = 1$ with probability p, and $\Delta_i = -1$ with probability $1 - p$ for all i. Thus, $E[\Delta_i] = 2p - 1$ holds for all i. Accordingly,

$$E(X_{n+1}|X_1, X_2, \ldots, X_n) = E(X_n + \Delta_{n+1}|X_1, X_2, \ldots, X_n)$$
$$= E(X_n|X_1, X_2, \ldots, X_n) + E(\Delta_{n+1}|X_1, X_2, \ldots, X_n)$$
$$= X_n + E[\Delta_{n+1}] = X_n + 2p - 1$$

Hence,

$$\text{If } \quad p \geq \frac{1}{2}, \quad E(X_{n+1}|X_1, X_2, \ldots, X_n) \geq X_n$$

will hold for $n \geq 1$, and a simple random walk will be a *submartingale*, and

$$\text{If} \quad p \le \frac{1}{2}, \quad E(X_{n+1}|X_1, X_2, \ldots, X_n) \le X_n$$

will hold for $n \ge 1$, and a simple random walk will be a *supermartingale*.

Definition 3 (*Doob type martingale*) Let $X_1, X_2, \ldots, X_n$ be an arbitrary sequence of random variables, and let $f(X) = f(X_1, X_2, \ldots, X_n)$ be a bounded function. If we define

$$Z_i = E(f(X)|X_1, X_2, \ldots, X_i)$$

then $\{Z_i; i \ge 1\}$ will be a special martingale, called as *Doob type martingale* with Z_i corresponding to $X_1, X_2, \ldots, X_i$.

Example 3 We toss a *fair coin* n times. Let

$$X_i = \begin{cases} 1 \text{ if } i\text{th toss results in heads} \\ 0 \text{ if } i\text{th toss results in tails} \end{cases}$$

Hence,

$$X_i \sim Bin\left(1, \frac{1}{2}\right) \quad \text{for } i = 1, 2, \ldots, n$$

and $f(X)$ function can be defined such that

$$f(X) = f(X_1, X_2, \ldots, X_n) = \sum_{i=1}^{n} X_i$$

is the total number of heads out of n fair tosses. Accordingly, $\{Z_i; i \ge 1\}$ will be a *Doob type martingale*, where

$$Z_0 = E(f(X_1, X_2, \ldots, X_n))$$

$$Z_i = E(f(X_1, X_2, \ldots, X_n)|X_1, X_2, \ldots, X_i) \text{ for } 1 \le i \le n$$

$$Z_n = E(f(X_1, X_2, \ldots, X_n)|X_1, X_2, \ldots, X_n) = f(X_1, X_2, \ldots, X_n)$$

17.2 Azuma-Hoeffding Inequality, Kolmogorov's Inequality, the Martingale Convergence Theorem

Definition 4 (*Azuma-Hoeffding inequality*) Let $\{Z_i; i \geq 1\}$ be a martingale, and let a constant $c_i > 0$ be for each Z_i such that

$$|Z_i - Z_{i-1}| \leq c_i$$

exists. Hence, for any $\lambda > 0, n \geq 1$,

$$P(|Z_n - Z_0| \geq \lambda) \leq 2exp\left\{-\frac{\lambda^2}{2\sum_{i=1}^{n} c_i^2}\right\}$$

will hold, and this inequality is called as *Azuma-Hoeffding inequality*.

Example 3 (revisited 1) Recall that

$$Z_i = E(f(X_1, X_2, \ldots, X_n)|X_1, X_2, \ldots, X_i) \text{ for } 1 \leq i \leq n$$

from Example 3. Accordingly,

$$Z_{i-1} = E(f(X_1, X_2, \ldots, X_n)|X_1, X_2, \ldots, X_{i-1})$$

and

$$
\begin{aligned}
|Z_i - Z_{i-1}| &= |E(f(X_1, X_2, \ldots, X_n)|X_1, X_2, \ldots, X_i) \\
&\quad -E(f(X_1, X_2, \ldots, X_n)|X_1, X_2, \ldots, X_{i-1})| \\
&= |(X_1 + X_2 + \cdots + X_i + E(X_{i+1} + X_{i+2} + \cdots + X_n)) \\
&\quad -(X_1 + X_2 + \cdots + X_{i-1} + E(X_i + X_{i+1} + \cdots + X_n))| \\
&= |X_i - E[X_i]| = \left|X_i - \frac{1}{2}\right| = \frac{1}{2} = c_i
\end{aligned}
$$

will hold for $i = 1, 2, \ldots, n$, since $X_i = 1$ or 0 for all i, and $E[X_i] = (1)(\frac{1}{2}) = \frac{1}{2}$ from Example 3. Recall

$$Z_0 = E(f(X_1, X_2, \ldots, X_n))$$

$$Z_n = f(X_1, X_2, \ldots, X_n)$$

from Example 3. Hence,

$$Z_0 = E(f(X_1, X_2, \ldots, X_n)) = E\left(\sum_{i=1}^{n} X_i\right) = \sum_{i=1}^{n} E[X_i] = n\left(\frac{1}{2}\right) = \frac{n}{2}$$

holds, and by Azuma-Hoeffding inequality for any $\lambda > 0, n \geq 1$,

$$P(|Z_n - Z_0| \geq \lambda) = P(|f(X_1, X_2, \ldots, X_n) - E(f(X_1, X_2, \ldots, X_n))| \geq \lambda)$$

$$\leq 2exp\left\{-\frac{\lambda^2}{2\sum_{i=1}^{n}\left(\frac{1}{2}\right)^2}\right\}$$

$$P\left(\left|f(X_1, X_2, \ldots, X_n) - \frac{n}{2}\right| \geq \lambda\right) \leq 2exp\left\{-\frac{\lambda^2}{\left(\frac{n}{2}\right)}\right\}$$

$$P\left(\left|f(X_1, X_2, \ldots, X_n) - \frac{n}{2}\right| \geq \lambda\right) \leq 2exp\left\{-\frac{2\lambda^2}{n}\right\}$$

Definition 5 (*Kolmogorov's inequality for submartingales and martingales*) Kolmogorov's inequality for submartingales, and an extension of Kolmogorov's inequality to martingales are defined in Table 17.2.

Remark 3 Note that the proof of Kolmogorov's inequality for submartingales follows mainly from the *Markov's inequality* provided in Table 7.8 in Chap. 7.

Example 2 (revisited 1) Recall that a *simple random walk* $\{X_n; n \geq 1\}$ is a submartingale if $p \geq \frac{1}{2}$. We consider

$$X_n = \Delta_1 + \Delta_2 + \cdots + \Delta_i + \cdots + \Delta_n$$

and assume that $p = 0.7$. Thus, as an example

$$P(\max(X_1, X_2, \ldots, X_{10}) > 5) \leq \frac{E[X_{10}]}{5}$$

holds by Kolmogorov's inequality for submartingales. By recalling that $E[\Delta_i] = 2p - 1$ for all i, for $p = 0.7$

$$E[X_{10}] = 10E[\Delta_i] = 10(2p - 1) = 10(2(0.7) - 1) = 4$$

Finally,

Table 17.2 Kolmogorov's inequality for submartingales and martingales

Kolmogorov's inequality for submartingales	Extension of Kolmogorov's inequality to martingales
If $\{Z_n; n \geq 1\}$ is a *nonnegative submartingale*, then for any $k > 0$, $P(\max(Z_1, Z_2, \ldots, Z_n) > k) \leq \frac{E[Z_n]}{k}$	If $\{Z_n; n \geq 1\}$ is a *martingale*, then for any $k > 0$, $P(\max(|Z_1|, |Z_2|, \ldots, |Z_n|) > k) \leq \frac{E[|Z_n|]}{k}$
	$P(\max(|Z_1|, |Z_2|, \ldots, |Z_n|) > k) \leq \frac{E[Z_n^2]}{k^2}$

$$P(\max(X_1, X_2, \ldots, X_{10}) > 5) \le \frac{E[X_{10}]}{5} = \frac{4}{5} = 0.8$$

Definition 6 (*The Martingale convergence theorem*) If $\{Z_n; n \ge 1\}$ is a martingale such that for some $M < \infty$,

$$E[|Z_n|] \le M \text{ for all } n$$

holds, then $\lim_{n \to \infty} Z_n$ exists with probability 1, and it is finite.

Problems

1. Show that a standard Brownian motion $\{B(t); t \ge 0\}$ is a martingale.

Solution

$$
\begin{aligned}
E(B(t+s)|B(u); 0 \le u \le s) &= E(B(s) + B(t+s) - B(s)|B(u); 0 \le u \le s) \\
&= E(B(s) + B(t+s) - B(s)|B(s)) \\
&= E(B(s)|B(s)) + E(B(t+s) - B(s)|B(s)) \\
&= B(s) + E(B(t+s) - B(s)) \\
&= B(s) + E(B(t)) \\
&= B(s) + 0 = B(s)
\end{aligned}
$$

Remark 1 Note that the second equality follows since a Brownian motion has *Markovian property*. Note also that $E(B(t+s) - B(s)|B(s)) = E(B(t+s) - B(s))$ due to the *independent increments*, and $E(B(t+s) - B(s)) = E(B(t))$ due to the *stationary increments* of a Brownian motion.

2. Let $\{N(t); t \ge 0\}$ be a homogeneous Poisson process with rate λ. Show that

$$\{Z(t); t \ge 0\}$$

is a martingale, where

$$Z(t) = N(t) - \lambda t$$

Solution

For $t > s$, we should show that

$$E(Z(t)|Z(s), Z(u); 0 \le u < s) = Z(s)$$

holds.

$$E(Z(t)|Z(s), Z(u); 0 \le u < s) = E(N(t) - \lambda t|N(s) - \lambda s, Z(u); 0 \le u < s)$$

$$\begin{aligned}
&= E(N(s) - \lambda s + N(t - s) - \lambda(t - s)|N(s) - \lambda s, \\
&\quad Z(u); 0 \leq u < s) \\
&= E(N(s) - \lambda s|N(s) - \lambda s, Z(u); 0 \leq u < s) \\
&\quad + E(N(t - s) - \lambda(t - s)|N(s) - \lambda s, \\
&\quad Z(u); 0 \leq u < s) \\
&= N(s) - \lambda s + E(N(t - s) - \lambda(t - s)) \\
&= N(s) - \lambda s + 0 = Z(s)
\end{aligned}$$

Remark 5 Note that $N(t) - \lambda t = N(s) - \lambda s + N(t - s) - \lambda(t - s)$ holds due to the *stationary increments*, and $E(N(t - s) - \lambda(t - s)|N(s) - \lambda s, Z(u); 0 \leq u < s) = E(N(t - s) - \lambda(t - s))$ due to the *independent increments* of a homogeneous Poisson process. Note also that $E[N(t - s)] = \lambda(t - s)$.

Remark 6 $\{Z(t); t \geq 0\}$ is a special Poisson process called as *compensated Poisson process*.

Chapter 18
Basics of Reliability Theory

Abstract This chapter has been devoted to the reliability theory. In the first part of the chapter, some basic definitions for a nonrepairable item including the time to failure, availability, reliability function, failure rate function, mean time to failure, and mean residual life have been defined, and the concepts have been made clear with some illustrative examples. In the second part, some basic concepts for a repairable item including the mean up time, mean down time, mean time to repair, mean time between failures, and average availability have been presented with the basic definitions and some examples. In the third part, the basic concepts for the systems with independent components including the state vector, probability vector, structure function, and availability function have been defined for different structures with some examples. Finally, in the fourth part, the positive dependency and negative dependency have been defined for the systems with dependent components, and two methods, i.e. square-root method, and β-factor model have been introduced, and illustrated with some examples. The problems at the end of chapter were aimed to enhance the comprehension of very important basic concepts in reliability theory.

Reliability theory is closely related to the basics of probability, and the basics of stochastic processes, especially related to the renewal processes. In this chapter, an item is considered as *nonrepairable* or *repairable*, and a system is considered with *independent* components or *dependent* components. Some basic parameters defined for the random variables such as the probability density function, cumulative distribution function, and expected value provided in Chap. 4 are adapted to the basic definitions of a nonrepairable item. The idea of the *alternating renewal process* introduced in Chap. 11 is adapted to the definition of the average availability for a repairable item. The formula for *independent events* provided in Chap. 3 is significant for the structure function and availability function of a *system with independent components*, while the *conditional probability* formula given in Chap. 3 should be recalled for the positive dependency/negative dependency, and the *square root method* for a *system with dependent components*. Several examples and problems with different structures illustrate the concepts defined in this chapter.

© Springer Nature Switzerland AG 2019

E. Bas, *Basics of Probability and Stochastic Processes*,
https://doi.org/10.1007/978-3-030-32323-3_18

18.1   Basic Definitions for a Nonrepairable Item

Definition 1 (*Time to failure*) In reliability theory, the *time to failure* T of a *nonrepairable* item is a random variable that describes the time from the *first operation* of the item to the *first failure*, and can be illustrated as in Fig. 18.1.

Remark 1 We may also define the *time to failure* for a *repairable item*, since we may not be interested in the processes after the first failure.

Definition 2 (Availability) The *availability* of a *nonrepairable item* can be defined as the *probability* that the item is functioning at time t, i.e.

$$A(t) = P(\text{item is functioning at time } t)$$

Definition 3 (*Reliability (survival) function*)

$$R(t) = P(T > t) = \overline{F}(t) = \int_{t}^{\infty} f(t)dt$$

can be defined as a *reliability (survival) function* for a *nonrepairable item* with the time to failure T, where $F(t)$ is the cumulative distribution function and $f(t)$ is the probability density function of T. The graph of a typical reliability function is depicted in Fig. 18.2.

Remark 2 Reliability function $R(t)$ describes the probability that a *nonrepairable item* will *survive* up to time t.

Definition 4 (*Failure rate function, hazard rate function*)

$$z(t) = \frac{f(t)}{R(t)}$$

can be defined as a *failure rate function* (*hazard rate function*) of a *nonrepairable item* with the time to failure T, where $f(t)$ is the probability density function of T, and $R(t)$ is the reliability (survival) function.

Remark 3 Failure rate function $z(t)$ basically means the *probability* that a *nonrepairable item will not function* any more given that it *survives up to time* t. In reliability theory, the term *failure rate function* is more commonly used rather than *hazard rate function*.

Fig. 18.1 Time to failure

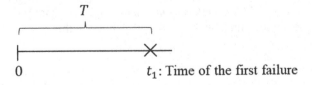

T

0 t_1: Time of the first failure

Fig. 18.2 Reliability (survival) function

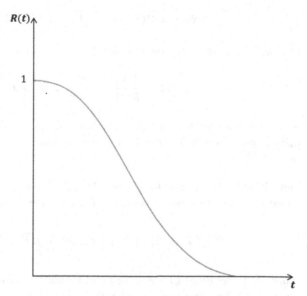

Definition 5 (*Increasing Failure Rate—IFR, Decreasing Failure Rate—DFR*) If a failure rate function $z(t)$ is an *increasing* function of t, then $F(t)$ corresponding to T is called as an *Increasing Failure Rate* (IFR) distribution. Similarly, if a failure rate function $z(t)$ is a *decreasing* function of t, then $F(t)$ corresponding to T is called as a *Decreasing Failure Rate* (DFR) distribution.

Remark 4 There is a special failure rate curve called as *bathtub curve* with three distinctive parts: A *DFR* part, a constant failure rate (*CFR*) part, and an *IFR* part. A bathtub curve is denoted in Fig. 18.3.

Example 1 Suppose that the time to failure T of an item is assumed to be an *exponential* random variable with rate λ. Hence, the reliability function of T will be

Fig. 18.3 Bathtub curve

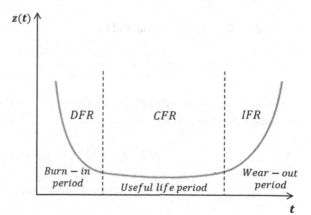

$$R(t) = P(T > t) = \overline{F}(t) = e^{-\lambda t} \quad \text{for} \quad t \geq 0$$

and the failure rate function of T will be

$$z(t) = \frac{f(t)}{R(t)} = \frac{\lambda e^{-\lambda t}}{e^{-\lambda t}} = \lambda \quad \text{for} \quad t \geq 0$$

Remark 5 Exponential random variable is a special random variable with a *constant failure rate function*. Thus, it is described as a random variable which is *both IFR and DFR*.

Definition 6 (*Mean time to failure—MTTF*) The mean time to failure ($MTTF$) of a *nonrepairable item* with the time to failure T can be defined as

$$MTTF = E[T] = \int_0^\infty t\, f(t)dt = \int_0^\infty \overline{F}(t)dt = \int_0^\infty R(t)dt$$

Example 1 (revisited 1) Since the time to failure T of the item is assumed to be an *exponential* random variable with rate λ, the mean time to failure will be

$$MTTF = E[T] = \int_0^\infty R(t)dt = \int_0^\infty e^{-\lambda t}dt = \frac{1}{\lambda}$$

Example 2 Consider the reliability function $R(t) = e^{-t}(t + 1)$ for a *nonrepairable item* with the time to failure T. Hence, the cumulative distribution function of T will be

$$F(t) = 1 - R(t) = 1 - e^{-t}(t + 1)$$

the probability density function of T will be

$$f(t) = \frac{dF(t)}{dt} = \frac{d(1 - R(t))}{dt} = -\frac{dR(t)}{dt} = e^{-t}(t + 1) - e^{-t} = te^{-t}$$

and the failure rate function will be

$$z(t) = \frac{f(t)}{R(t)} = \frac{te^{-t}}{e^{-t}(t + 1)} = \frac{t}{t + 1} = 1 - \frac{1}{t + 1}$$

Since $z(t)$ is an *increasing* function of t, $F(t)$ is an IFR distribution. Additionally, the mean time to failure will be

$$MTTF = E[T] = \int_0^\infty R(t)dt$$

$$= \int_0^\infty e^{-t}(t + 1)dt = 2$$

Definition 7 (Mean residual life—*MRL*) Let

$$R(x|t) = P(\{T > x + t\}|\{T > t\})$$

be the conditional reliability function of an item that survives up to time t. Accordingly, the mean residual life (MRL) of an item that survives up to time t can be defined as

$$
\begin{aligned}
MRL(t) &= \int_0^\infty R(x|t)dx = \int_0^\infty P(\{T > x + t\}|\{T > t\})dx \\
&= \int_0^\infty \frac{P(\{T > x + t\} \cap \{T > t\})}{P(T > t)}dx \\
&= \int_0^\infty \frac{P(T > x + t)}{P(T > t)}dx = \int_0^\infty \frac{R(x + t)}{R(t)}dx \\
&= \frac{1}{R(t)}\int_0^\infty R(x + t)dx = \frac{1}{R(t)}\int_t^\infty R(x)dx
\end{aligned}
$$

Remark 6 Note that $MRL(0) = \int_0^\infty R(x|0)dx = \int_0^\infty R(x)dx = MTTF$

Example 1 (revisited 2) The mean residual life for an item with *exponentially* distributed time to failure T will be

$$
\begin{aligned}
MRL(t) &= \frac{1}{R(t)}\int_t^\infty R(x)dx \\
&= \frac{1}{e^{-\lambda t}}\int_t^\infty e^{-\lambda x}dx = \frac{1}{e^{-\lambda t}}\left(\frac{1}{\lambda}e^{-\lambda t}\right) = \frac{1}{\lambda} = MTTF
\end{aligned}
$$

Remark 7 An exponential random variable is a special random variable with $MRL(t) = MTTF$ for all $t \geq 0$ due its unique *memoryless property*.

Example 2 (revisited 1) The mean residual life of an item with reliability function $R(t) = e^{-t}(t + 1)$ will be

$$
\begin{aligned}
MRL(t) &= \frac{1}{R(t)}\int_t^\infty R(x)dx \\
&= \frac{1}{e^{-t}(t + 1)}\int_t^\infty e^{-x}(x + 1)dx \\
&= \frac{1}{e^{-t}(t + 1)}e^{-t}(t + 2) = \frac{(t + 2)}{(t + 1)}
\end{aligned}
$$

18.2   Basic Definitions for a Repairable Item

Definition 8 (*Mean up time—MUT*) The mean up time (MUT) is the mean time during which *a repairable item* is *functioning*.

Remark 8 In case of a *nonrepairable item*, MUT is equal to $MTTF$.

Definition 9 (Mean down time—*MDT*) The mean down time (MDT) is the mean time during which *a repairable item* is *not functioning*.

Remark 9 Several probability distributions can be considered for the *down time*. However, *lognormal distribution* is one of the most frequently used distributions.

Definition 10 (*Mean time to repair—MTTR*) The mean time to repair ($MTTR$) is the mean time during which *a repairable item* is *being repaired*.

Remark 10 Note that $MDT \geq MTTR$ is assumed to hold due to several reasons such as waiting for the spare parts or the repair personnel, or additional time needed to put the item into operation after the repair.

Definition 11 (*Mean time between failures—MTBF*) The mean time between failures ($MTBF$) can be defined for *a repairable item* as the mean time from *one failure* to the *next failure*, that is

$$MTBF = MUT + MDT$$

$MUT, MDT, MTTR$, and $MTBF$ are shown on a $X(t)$ function given in Fig. 18.4.

Note that $X(t)$ is a function such that

$$X(t) = \begin{cases} 1 \; if \; item \; is \; up \; at \; time \; t \\ 0 \; if \; item \; is \; down \; at \; time \; t \end{cases}$$

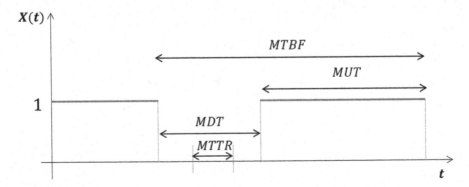

Fig. 18.4 $X(t)$ function with $MUT, MDT, MTTR, MTBF$ (System Reliability Theory, Models, Statistical Methods, and Applications by Rausand and Hoyland.)

Definition 12 (*Average availability*) The *average availability* for *a repairable item* can be defined as

$$A_{av} = \frac{MUT}{MUT + MDT} = \frac{MUT}{MTBF}$$

Remark 11 *Average availability* for *a repairable item* is analogous to the *limiting probability that the system is on* in an *alternating renewal process* provided in Definition 9 of Chap. 11.

Example 3 We consider a *repairable* item. The *up time* of this item is *exponentially* distributed with rate $\frac{1}{4}$ per month. When the item fails, time to find the spare parts and the repair personnel is *uniformly* distributed over $(1, 3)$ weeks, and it takes *exponentially* distributed time with rate $\frac{1}{8}$ per hour to fix the item. After the item is fixed, it is put into operation immediately.

If we assume that the item is working 30 days a month and 24 h a day, then the *mean up time* (MUT) of the item will be

$$MUT = \frac{1}{\left(\frac{1}{4}\right)} = 4\,\text{months} = (4)(30)(24) = 2,880\,\text{h}$$

The *mean time to repair* $(MTTR)$ of the item will be

$$MTTR = \frac{1}{\left(\frac{1}{8}\right)} = 8\,\text{h}$$

If we assume that we search for the spare parts and the repair personnel every day and 24 h a day, then the *mean down time* (MDT) of the item will be

$$MDT = \left(\frac{1+3}{2}\right)(7)(24) + \frac{1}{\left(\frac{1}{8}\right)} = 344\,\text{h}$$

Accordingly, the *mean time between failures* $(MTBF)$ will be

$$MTBF = MUT + MDT = 2,880 + 344 = 3,224\,\text{h}$$

Finally, the *average availability* of the item will be

$$A_{av} = \frac{MUT}{MUT + MDT} = \frac{MUT}{MTBF} = \frac{2,880}{3,224} = 0.8933$$

18.3 Systems with Independent Components

Definition 13 (*State vector vs. probability vector for a system with independent components*) For a system with n *independent* components, the *state vector* can be defined as

$$x = (x_1, x_2, \ldots, x_n)$$

where

$$x_i = \begin{cases} 1 \text{ if component } i \text{ is up} \\ 0 \text{ if component } i \text{ is down} \end{cases}$$

for $i = 1, 2, 3, \ldots, n$. When we consider the entries of the state vector as the random variables such that

$$X = (X_1, X_2, \ldots, X_n)$$

where

$$X_i = \begin{cases} 1 \text{ if component } i \text{ is up} \\ 0 \text{ if component } i \text{ is down} \end{cases}$$

then the *probability vector* should be defined as $p = (p_1, p_2, \ldots, p_n), i = 1, 2, 3, \ldots, n$, where

p_i : The probability that component i is up (Availability of component i)

Example 4 The *state vector* for a system with four components, in which only component 1 and component 4 are up, will be

$$x = (1, 0, 0, 1)$$

where $x_1 = 1, x_2 = 0, x_3 = 0, x_4 = 1$. The *probability vector* for this system could be defined as

$$p = (0.8, 0.2, 0.4, 0.9)$$

where $p_1 = 0.8, p_2 = 0.2, p_3 = 0.4, p_4 = 0.9$.

Definition 14 (*Structure function vs. availability function for a system with independent components*) The definitions of the structure function and the availability function, and the formulations for some special structures are provided in Table 18.1.

Table 18.1 Structure function versus availability function

Structure function (definition)	Availability function (definition)
The structure function can be defined for a system such that $$\phi(x) = \begin{cases} 1 & \text{if system is up for } x \\ 0 & \text{if system is down for } x \end{cases}$$ where $x = (x_1, x_2, \ldots, x_n)$ is the *state vector*.	Availability function can be defined for a system such that $A = P(\phi(X) = 1)$ where $$\phi(X) = \begin{cases} 1 & \text{if system is up for } X \\ 0 & \text{if system is down for } X \end{cases}$$ and $X = (X_1, X_2, \ldots, X_n)$ is a vector with the states as the *random variables*. In case of *independent components*, $A = A(p)$ can be denoted, where $p = (p_1, p_2, \ldots, p_n)$ is the *probability vector*.
Structure function for a series structure	Availability function for a series structure with independent components
$\phi(x) = \min(x_1, x_2, \ldots, x_n)$ $$= \prod_{i=1}^{n} x_i$$ 1　　2　　3　　...　　n	$$A(p) = \prod_{i=1}^{n} p_i$$
Structure function for a parallel structure	Availability function for a parallel structure with independent components
$\phi(x) = \max(x_1, x_2, \ldots, x_n)$ $$= 1 - \prod_{i=1}^{n}(1 - x_i)$$ 1 2 3 n	$$A(p) = 1 - \prod_{i=1}^{n}(1 - p_i)$$
Structure function for a k-out of-n structure	Availability function for a k-out of-n structure with independent components
$$\phi(x) = \begin{cases} 1 \text{ if } \sum_{i=1}^{n} x_i \geq k \\ 0 \text{ if } \sum_{i=1}^{n} x_i < k \end{cases}$$	$$A(p) = P\left(\sum_{i=1}^{n} X_i \geq k\right)$$ *Special case* If $p_1 = p_2 = \cdots = p_n = p$, then $$A(p) = \sum_{i=k}^{n} \binom{n}{i} p^i (1 - p)^{n-i}$$

Example 5 Let the probability vector for a system with *four independent* components
be

$$p = (p_1, p_2, p_3, p_4)$$

If the system is assumed to have a *series structure*, then the *availability function*
will be

$$A(p) = p_1 p_2 p_3 p_4$$

If the system is assumed to have a *parallel structure*, then the *availability function*
will be

$$A(p) = 1 - [(1 - p_1)(1 - p_2)(1 - p_3)(1 - p_4)]$$
$$= p_1 + p_2 + p_3 + p_4 - p_1 p_2 - p_1 p_3 - p_1 p_4 - p_2 p_3 - p_2 p_4 - p_3 p_4$$
$$+ p_1 p_2 p_3 + p_1 p_2 p_4 + p_1 p_3 p_4 + p_2 p_3 p_4 - p_1 p_2 p_3 p_4$$

If the system is assumed to have a 3—*out of*—4 *structure*, then the *availability
function* will be

$$A(p) = p_1 p_2 p_3 (1 - p_4) + p_1 p_2 (1 - p_3) p_4 + p_1 (1 - p_2) p_3 p_4$$
$$+ (1 - p_1) p_2 p_3 p_4 + p_1 p_2 p_3 p_4$$
$$= p_1 p_2 p_3 + p_1 p_2 p_4 + p_1 p_3 p_4 + p_2 p_3 p_4 - 3 p_1 p_2 p_3 p_4$$

If the system is assumed to have the structure in Fig. 18.5, then the *availability
function* will be

$$A(p) = [1 - (1 - p_1)(1 - p_2)][1 - (1 - p_3)(1 - p_4)]$$
$$= (p_1 + p_2 - p_1 p_2)(p_3 + p_4 - p_3 p_4)$$
$$= p_1 p_3 + p_1 p_4 + p_2 p_3 + p_2 p_4 - p_1 p_2 p_3 - p_1 p_2 p_4$$
$$- p_1 p_3 p_4 - p_2 p_3 p_4 + p_1 p_2 p_3 p_4$$

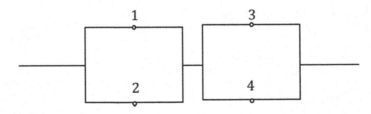

Fig. 18.5 Structure 1 for example 5

Fig. 18.6 Structure 2 for
example 5

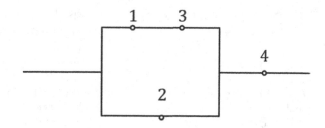

If the system is assumed to have the structure in Fig. 18.6, then the *availability function* will be

$$A(p) = [1 - (1 - p_1 p_3)(1 - p_2)]p_4$$
$$= p_2 p_4 + p_1 p_3 p_4 - p_1 p_2 p_3 p_4$$

Definition 15 (*Path set, minimal path set, minimal path vector vs. cut set, minimal cut set, minimal cut vector for a system with independent components*) The path set, minimal path set, minimal path vector; and the cut set, minimal cut set, and minimal cut vector can defined as given in Table 18.2 for a system with independent components.

Example 6 If a system is assumed to have the structure in Fig. 18.5, then the *minimum path sets* will be $\{1, 3\}, \{1, 4\}, \{2, 3\}, \{2, 4\}$, and the *minimum cut sets* will be $\{1, 2\}, \{3, 4\}$. Accordingly, the minimum path vectors will be $(1, 0, 1, 0), (1, 0, 0, 1), (0, 1, 1, 0), (0, 1, 0, 1)$, and the minimum cut vectors will be $(0, 0, 1, 1), (1, 1, 0, 0)$.

If a system is assumed to have the structure in Fig. 18.6, then the *minimum path sets* will be $\{1, 3, 4\}, \{2, 4\}$, and the *minimum cut sets* will be $\{1, 2\}, \{2, 3\}, \{4\}$. Accordingly, the minimum path vectors will be $(1, 0, 1, 1), (0, 1, 0, 1)$, and the minimum cut vectors will be $(0, 0, 1, 1), (1, 0, 0, 1), (1, 1, 1, 0)$.

Table 18.2 Path set, minimal path set, minimal path vector versus cut set, minimal cut set, minimal cut vector

Path set	Cut set
The set with the components whose *functioning* ensures the *functioning* of the system.	The set with the components whose *failure* ensures the *failure* of the system.
Minimal path set	*Minimal cut set*
The set B with the minimal components whose *functioning* ensures the *functioning* of the system.	The set C with the minimal components whose *failure* ensures the *failure* of the system.
Minimal path vector	*Minimal cut vector*
The state vector with the elements $x_i = 1, i \in B$, and $x_j = 0, j \notin B$	The state vector with the elements $x_i = 0, i \in C$, and $x_j = 1, j \notin C$

Table 18.3 Positive dependency versus negative dependency

Positive dependency	Negative dependency
If $P(\overline{C}_1\|\overline{C}_2) \geq P(\overline{C}_1)$ and $P(\overline{C}_2\|\overline{C}_1) \geq P(\overline{C}_2)$ hold, then $P(\overline{C}_1 \cap \overline{C}_2) \geq P(\overline{C}_1)P(\overline{C}_2)$ Hence, component 1 and component 2 are said to be *positively dependent*.	If $P(\overline{C}_1\|\overline{C}_2) \leq P(\overline{C}_1)$ and $P(\overline{C}_2\|\overline{C}_1) \leq P(\overline{C}_2)$ hold, then $P(\overline{C}_1 \cap \overline{C}_2) \leq P(\overline{C}_1)P(\overline{C}_2)$ Hence, component 1 and component 2 are said to be *negatively dependent*.

18.4 Systems with Dependent Components

Definition 16 (*Positive dependency vs. negative dependency*) We consider a system with *two dependent* components. Let

$$\overline{C}_i : \text{ Event that component } i \text{ fails, } \quad i = 1, 2$$

Hence, the positive dependency and negative dependency of these dependent components can be defined as in Table 18.3.

Remark 12 The condition for the *positive dependency* can be generalized as

$$P(\overline{C}_1 \cap \overline{C}_2 \cap \ldots \cap \overline{C}_n) \geq P(\overline{C}_1)P(\overline{C}_2)\ldots P(\overline{C}_n) \quad \text{for} \quad n \geq 2$$

and the condition for the *negative dependency* can be generalized as

$$P(\overline{C}_1 \cap \overline{C}_2 \cap \ldots \cap \overline{C}_n) \leq P(\overline{C}_1)P(\overline{C}_2)\ldots P(\overline{C}_n) \quad \text{for} \quad n \geq 2$$

Definition 17 (*Square root method*) We consider a *parallel system* with *two components*. Let

$$\overline{C}_i: \text{Event that component } i \text{ fails, } \quad i = 1, 2$$

Hence,

$$P(\overline{C}_1 \cap \overline{C}_2): \text{Probability that the parallel system fails}$$

It is clear that

$$P(\overline{C}_1 \cap \overline{C}_2) \leq P(\overline{C}_i) \quad i = 1, 2$$

Accordingly,

$$P(\overline{C}_1 \cap \overline{C}_2) \leq \min(P(\overline{C}_1), P(\overline{C}_2))$$

If we assume that the events $\overline{C}_1$ and $\overline{C}_2$ are *positively dependent*, then

$$P(\overline{C}_1 \cap \overline{C}_2) \geq P(\overline{C}_1)P(\overline{C}_2)$$

from Table 18.3. Finally,

$$P(\overline{C}_1)P(\overline{C}_2) \leq P(\overline{C}_1 \cap \overline{C}_2) \leq \min(P(\overline{C}_1), P(\overline{C}_2))$$

and $P(\overline{C}_1 \cap \overline{C}_2)$ can be approximated as

$$P(\overline{C}_1 \cap \overline{C}_2) \cong \sqrt{[P(\overline{C}_1)P(\overline{C}_2)][\min(P(\overline{C}_1), P(\overline{C}_2))]}$$

which is called as the *square root method*.

Remark 13 Note that the *square root method* can be generalized as

$$P(\overline{C}_1 \cap \overline{C}_2 \cap \ldots \cap \overline{C}_n) \cong \sqrt{[P(\overline{C}_1)P(\overline{C}_2)\ldots P(\overline{C}_n)][\min(P(\overline{C}_1), P(\overline{C}_2), \ldots, P(\overline{C}_n))]}$$

for $n \geq 2$.

Example 7 Consider the parallel system in Fig. 18.7 with four *positively dependent* components.
Let

$$P(\overline{C}_1) = 0.3$$
$$P(\overline{C}_2) = 0.8$$
$$P(\overline{C}_3) = 0.6$$
$$P(\overline{C}_4) = 0.5$$

By using the *square root method*, the probability that this parallel system fails will be

Fig. 18.7 Structure for example 7

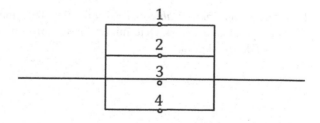

$$P(\overline{C}_1 \cap \overline{C}_2 \cap \overline{C}_3 \cap \overline{C}_4)$$
$$\cong \sqrt{[P(\overline{C}_1)P(\overline{C}_2)P(\overline{C}_3)P(\overline{C}_4)][\min(P(\overline{C}_1), P(\overline{C}_2), P(\overline{C}_3), P(\overline{C}_4))]}$$
$$\cong \sqrt{[(0.3)(0.8)(0.6)(0.5)][\min(0.3, 0.8, 0.6, 0.5)]}$$
$$\cong \sqrt{(0.072)(0.3)} = 0.1470$$

Definition 18 (β-factor model) We consider a system with n identical components. Let λ^i be the *rate of individual* failures of a component, let λ_{CCF} be the *rate of common cause failures* (CCF), and let λ be the *total failure rate* of a component. Hence,

$$\beta = \frac{\lambda_{CCF}}{\lambda^i + \lambda_{CCF}} = \frac{\lambda_{CCF}}{\lambda}$$

can be defined, where β refers to the relative portion of CCF among all failures.

Remark 14 Note that CCF can be defined as a specific type of single shared failure that causes several other failures.

Example 8 Consider a system with *two identical* components, and consider β for the relative portion of CCF among all failures. Recall that

$$\beta = \frac{\lambda_{CCF}}{\lambda^i + \lambda_{CCF}} = \frac{\lambda_{CCF}}{\lambda}$$

Accordingly, the rate of CCF will be

$$\lambda_{CCF} = \beta\lambda$$

and the rate of *individual* failures for one component will be

$$\lambda^i = \lambda(1 - \beta)$$

As a result, the rate of individual failures for *two* components will be

$$2\lambda(1 - \beta)$$

Problems

1. Consider a *nonrepairable item* with the time to failure T uniformly distributed over the interval (α, β). Determine whether uniform random variable is an IFR or DFR distribution.

Solution

Since

$$T \sim Uniform(\alpha, \beta)$$

the probability density function of T will be

$$f(t) = \frac{1}{\beta - \alpha}$$

the cumulative distribution function of T will be

$$F(t) = \frac{t - \alpha}{\beta - \alpha} \quad \text{for } \alpha < t < \beta$$

and the *reliability function* will be

$$R(t) = 1 - F(t) = \frac{\beta - t}{\beta - \alpha} \quad \text{for } \alpha < t < \beta$$

Finally, the *failure rate function* will be

$$z(t) = \frac{f(t)}{R(t)} = \frac{\left(\frac{1}{\beta - \alpha}\right)}{\left(\frac{\beta - t}{\beta - \alpha}\right)} = \frac{1}{\beta - t} \quad \text{for } \alpha < t < \beta$$

Since $z(t)$ is an *increasing* function of t, a uniform random variable is an IFR distribution.

2. Consider the structure in Fig. 18.8 for a system with independent components.

Determine the structure function and the availability function for this system.

Solution

The *structure function* for this system will be

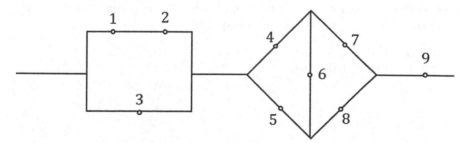

Fig. 18.8 Structure for problem 2

Fig. 18.9 Bridge system for
problem 2

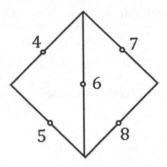

$$\phi(x) = [1 - (1 - x_1 x_2)(1 - x_3)]$$
$$[1 - (1 - x_4 x_7)(1 - x_4 x_6 x_8)(1 - x_5 x_8)(1 - x_5 x_6 x_7)]x_9$$

and the *availability function* for this system will be

$$A(p) = [1 - (1 - p_1 p_2)(1 - p_3)]$$
$$[1 - (1 - p_4 p_7)(1 - p_4 p_6 p_8)(1 - p_5 p_8)(1 - p_5 p_6 p_7)]p_9$$

Remark 15 The part of the system in Fig. 18.9 is called as a *Bridge system.*
   We note that the *minimum path sets* for this bridge system are
$\{4, 7\}$, $\{4, 6, 8\}$, $\{5, 8\}$, $\{5, 6, 7\}$, and consider the *structure function* for this part of
the system as

$$\max(x_4 x_7, x_4 x_6 x_8, x_5 x_8, x_5 x_6 x_7)$$

and proceed with

$$\max(x_4 x_7, x_4 x_6 x_8, x_5 x_8, x_5 x_6 x_7)$$
$$= [1 - (1 - x_4 x_7)(1 - x_4 x_6 x_8)(1 - x_5 x_8)(1 - x_5 x_6 x_7)]$$

The availability function of this bridge system can be determined analogously.

3. Consider a 3—out of—4 structure with *independent components*, and the fol-
    lowing two probability vectors for this system:

    (a) $p = (0.4, 0.3, 0.7, 0.5)$
    (b) $p = (0.4, 0.4, 0.4, 0.4)$

Determine the availability function of this system by considering (a) and (b).

Solution

(a) The availability function with $p = (0.4, 0.3, 0.7, 0.5)$ will be

$$A(p) = P\left(\sum_{i=1}^{4} X_i \geq 3\right) = (0.4)(0.3)(0.7)(1 - 0.5)$$
$$+ (0.4)(0.3)(1 - 0.7)(0.5)$$
$$+ (0.4)(1 - 0.3)(0.7)(0.5)$$
$$+ (1 - 0.4)(0.3)(0.7)(0.5)$$
$$+ (0.4)(0.3)(0.7)(0.5) = 0.263$$

(b) The availability function with $p = (0.4, 0.4, 0.4, 0.4)$ will be

$$A(p) = P\left(\sum_{i=1}^{4} X_i \geq 3\right) = \binom{4}{3}(0.4)^3(0.6)^1 + \binom{4}{4}(0.4)^4(0.6)^0 = 0.1792$$

4. A *nonrepairable item* with the time to failure T has a *constant failure rate* for all $t \geq 0$. This item can function for at least 200 h without any failure with probability 0.6.

   (a) Find the failure rate function for this item.
   (b) Find the probability that the item will survive up to 800 h given that it functions for 300 h without any failure.
   (c) Find the mean time to failure of this item.
   (d) Find the mean residual life of this item at $t = 300$ h.
   (e) What is the probability that the item will fail at some time up to 1,500 h given that it survives up to 500 h?

Solution
Since the item has a *constant failure rate* for all $t \geq 0$,

$$T \sim exp(\lambda)$$

(a) Since this item can function for at least 200 h without any failure with probability 0.6,

$$P(T > 200) = R(200) = 0.6$$
$$e^{-\lambda 200} = 0.6$$
$$-\lambda 200 = \ln(0.6)$$

$$\lambda = -\frac{\ln(0.6)}{200} = 0.0026$$

Accordingly,

$$T \sim exp(0.0026)$$

and the failure rate function will be

$$z(t) = \lambda = 0.0026$$

(b) The probability that the item will survive up to 800 h given that it functions for 300 h without any failure will be

$$P(T > 800|T > 300) = P(T > 500)$$

by *memoryless property* of exponential random variable, and

$$P(T > 500) = R(500) = e^{-\lambda 500} = e^{-(0.0026)500} = 0.2725$$

(c) The mean time to failure of this item will be

$$MTTF = E[T] = \frac{1}{\lambda} = \frac{1}{0.0026} = 384.62 \, h$$

(d) The mean residual life of this item will be *constant* for any $t \geq 0$ such that

$$MRL(t) = MTTF = E[T] = \frac{1}{\lambda} = \frac{1}{0.0026} = 384.62 \, h$$

(e) The probability that the item will fail at some time up to 1,500 h given that it survives up to 500 h will be

$$
\begin{aligned}
P(T < 1,500|T > 500) &= \frac{P(500 < T < 1,500)}{P(T > 500)} \\
&= \frac{F(1,500) - F(500)}{R(500)} \\
&= \frac{\left(1 - e^{-(0.0026)1,500}\right) - \left(1 - e^{-(0.0026)500}\right)}{e^{-(0.0026)500}} = 0.9257
\end{aligned}
$$

5. Consider the structure in Fig. 18.10 for a system with independent components.

Determine the structure function, availability function, minimal path sets, and minimal cut sets for this system.

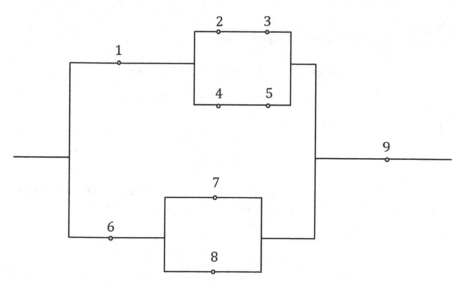

Fig. 18.10 Structure for problem 5

Solution

The *structure function* of this system will be

$$\phi(x) = [1 - (1 - x_1(1 - (1 - x_2x_3)(1 - x_4x_5)))(1 - x_6(1 - (1 - x_7)(1 - x_8)))]x_9$$

The *availability function* of this system will be

$$A(p) = [1 - (1 - p_1(1 - (1 - p_2p_3)(1 - p_4p_5)))(1 - p_6(1 - (1 - p_7)(1 - p_8)))]p_9$$

The *minimal path sets* of this system will be
$\{1, 2, 3, 9\}, \{1, 4, 5, 9\}, \{6, 7, 9\}, \{6, 8, 9\}$, and the *minimal cut sets* of this system will be $\{1, 6\}, \{2, 4, 6\}, \{3, 5, 6\}, \{1, 7, 8\}, \{2, 4, 7, 8\}, \{3, 5, 7, 8\}, \{9\}$.

6. A system with the structure given in Fig. 18.11 has to maintain a 99% availability.
   If each component of this system has the same availability *independent* from each
   other, what is the availability of each component?

Fig. 18.11 Structure for
problem 6

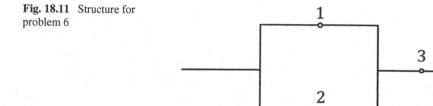

Solution

The availability function of the system will be

$$A(p) = [1 - (1 - p)(1 - p)]p = 0.99$$

p : Availability of each component

Hence,

$$p^3 - 2p^2 + 0.99 = 0$$

When we solve this cubic equation, we find the roots 0.99, 1.618 and -0.618. Thus, the availability of each component should also be $p = 0.99$.

Area Under the Standard Normal Curve
to the Left of z

© Springer Nature Switzerland AG 2019
E. Bas, *Basics of Probability and Stochastic Processes*,
https://doi.org/10.1007/978-3-030-32323-3

	.00	.01	.02	.03	.04	.05	.06	.07	.08	.09
.0	.5000	.5040	.5080	.5120	.5160	.5199	.5239	.5279	.5319	.5359
.1	.5398	.5438	.5478	.5517	.5557	.5596	.5636	.5675	.5714	.5753
.2	.5793	.5832	.5871	.5910	.5948	.5987	.6026	.6064	.6103	.6141
.3	.6179	.6217	.6255	.6293	.6331	.6368	.6406	.6443	.6480	.6517
.4	.6554	.6591	.6628	.6664	.6700	.6736	.6772	.6808	.6844	.6879
.5	.6915	.6950	.6985	.7019	.7054	.7088	.7123	.7157	.7190	.7224
.6	.7257	.7291	.7324	.7357	.7389	.7422	.7454	.7486	.7517	.7549
.7	.7580	.7611	.7642	.7673	.7704	.7734	.7764	.7794	.7823	.7852
.8	.7881	.7910	.7939	.7967	.7995	.8023	.8051	.8078	.8106	.8133
.9	.8159	.8186	.8212	.8238	.8264	.8289	.8315	.8340	.8365	.8389
1.0	.8413	.8438	.8461	.8485	.8508	.8531	.8554	.8577	.8599	.8621
1.1	.8643	.8665	.8686	.8708	.8729	.8749	.8770	.8790	.8810	.8830
1.2	.8849	.8869	.8888	.8907	.8925	.8944	.8962	.8980	.8997	.9015
1.3	.9032	.9049	.9066	.9082	.9099	.9115	.9131	.9147	.9162	.9177
1.4	.9192	.9207	.9222	.9236	.9251	.9265	.9279	.9292	.9306	.9319
1.5	.9332	.9345	.9357	.9370	.9382	.9394	.9406	.9418	.9429	.9441
1.6	.9452	.9463	.9474	.9484	.9495	.9505	.9515	.9525	.9535	.9545
1.7	.9554	.9564	.9573	.9582	.9591	.9599	.9608	.9616	.9625	.9633
1.8	.9641	.9649	.9656	.9664	.9671	.9678	.9686	.9693	.9699	.9706
1.9	.9713	.9719	.9726	.9732	.9738	.9744	.9750	.9756	.9761	.9767
2.0	.9772	.9778	.9783	.9788	.9793	.9798	.9803	.9808	.9812	.9817

(continued)

(continued)

	.00	.01	.02	.03	.04	.05	.06	.07	.08	.09
2.1	.9821	.9826	.9830	.9834	.9838	.9842	.9846	.9850	.9854	.9857
2.2	.9861	.9864	.9868	.9871	.9875	.9878	.9881	.9884	.9887	.9890
2.3	.9893	.9896	.9898	.9901	.9904	.9906	.9909	.9911	.9913	.9916
2.4	.9918	.9920	.9922	.9925	.9927	.9929	.9931	.9932	.9934	.9936
2.5	.9938	.9940	.9941	.9943	.9945	.9946	.9948	.9949	.9951	.9952
2.6	.9953	.9955	.9956	.9957	.9959	.9960	.9961	.9962	.9963	.9964
2.7	.9965	.9966	.9967	.9968	.9969	.9970	.9971	.9972	.9973	.9974
2.8	.9974	.9975	.9976	.9977	.9977	.9978	.9979	.9979	.9980	.9981
2.9	.9981	.9982	.9982	.9983	.9984	.9984	.9985	.9985	.9986	.9986
3.0	.9987	.9987	.9987	.9988	.9988	.9989	.9989	.9989	.9990	.9990
3.1	.9990	.9991	.9991	.9991	.9992	.9992	.9992	.9992	.9993	.9993
3.2	.9993	.9993	.9994	.9994	.9994	.9994	.9994	.9995	.9995	.9995
3.3	.9995	.9995	.9995	.9996	.9996	.9996	.9996	.9996	.9996	.9997
3.4	.9997	.9997	.9997	.9997	.9997	.9997	.9997	.9997	.9997	.9998

References

Bourier, G.: Wahrscheinlichkeitsrechnung und schliessende Statistik: Praxisorientierte Einführung. Mit Aufgaben und Lösungen, Springer Fachmedien Wiesbaden (2013)

Cinlar, E.: Introduction to Stochastic Processes. Dover Publications, Inc. (1975)

Doob, J.L.: Stochastic Processes. Wiley Classics Library (1953)

Durrett, R.: Essentials of Stochastic Processes. Springer Texts in Statistics, 2nd edn (2012)

Feinberg, E.A., Shwartz, A.: Handbook of Markov Decision Processes, Methods and Applications. Kluwer's International Series in Operations Research & Management Science (2002)

Gass, S.I., Fu, M.C. (eds.): Kendall's Notation. Encyclopedia of Operations Research and Management Science. Springer (2013)

Karlin, S., Taylor, H.: A First Course in Stochastic Processes, 2nd edn. Academic Press, Elsevier (1975)

Marengo, J.E., Farnsworth, D.L., Stefanic, L.: A geometric derivation of the Irwin-Hall distribution. Int. J. Math. Math. Sci. **2017**, Article ID 3571419

Pinsky, M.A., Karlin, S.: An Introduction to Stochastic Modeling, 4th edn. Academic Press, Elsevier (2011)

Pitman, J.: Probability. Springer (1993)

Puterman, M.L.: Markov Decision Processes: Discrete Stochastic Dynamic Programming. Wiley Series in Probability and Statistics (2005)

Rausand, M., Hoyland, A.: System Reliability Theory, Models, Statistical Methods, and Applications, 2nd edn. Wiley-Interscience (2004)

Ross, S.: A First Course in Probability, 8th edn. Prentice Hall, Pearson (2010)

Ross, S.M.: Introduction to Probability Models, 10th edn. Academic Press, Elsevier (2010)

Ross, S.M.: Stochastic Processes, 2nd edn. Wiley (1996)

Shmueli, G.: Practical Acceptance Sampling. Axelrod Schnall Publishers (2016)

Sigman, K., Wolff, R.W.: A review of regenerative processes. SIAM Rev. Publ. Soc. Ind. Appl. Math. **35**, 269–288 (1993)

Sigman, K.: Columbia University, Lecture Notes on Stochastic Modelling I. http://www.columbia.edu/~ks20/stochastic-I/stochastic-I.html

Sigman, K.: Columbia University, Lecture Notes on Brownian Motion. http://www.columbia.edu/~ks20/FE-Notes/4700-07-Notes-BM.pdf

Sigman, K.: Columbia University, Lecture Notes on Poisson Processes, and Compound (Batch) Processes. http://www.columbia.edu/~ks20/4703-Sigman/4703-07-Notes-PP-NSPP.pdf

© Springer Nature Switzerland AG 2019
E. Bas, *Basics of Probability and Stochastic Processes*,
https://doi.org/10.1007/978-3-030-32323-3

Index

A

Absorbing state, 185, 186, 218
Acceptance sampling, 55, 66
Alternating renewal process, 163, 170, 171, 178, 273, 279
Aperiodic state, 187, 197, 198
Approximated probability, 90, 111, 118, 160
Approximation, 56, 71, 79, 80, 90, 93, 117, 149, 154, 155
Arrival time, 81, 131–133, 138, 145, 149, 156
Availability
   average availability, 273, 279
Availability function
   k-out of-n structure, 281
   parallel structure, 281, 282
   series structure, 281, 282
Average age, 167, 168
Average excess, 167, 168, 175
Average spread, 167
Axioms of probability, 15, 16
Azuma-Hoeffding inequality, 265, 269, 270

B

Backwards recurrence time, 167
Balance equations, 189, 190, 207, 217, 233, 235–239, 242, 243, 245–250
Bank, 71, 88, 233, 234, 241, 242
Basic principle of counting, 3, 4, 23, 28
Bathtub curve
   constant failure rate (CFR), 275
   decreasing failure rate (DFR), 275
   increasing failure rate (IFR), 275
Bayes' formula, 27, 30, 31
Bernoulli random variable, 55, 60

Binomial random variable, 55–58, 60, 63, 66, 68, 71, 79, 80, 90, 93, 99, 100, 113, 149, 154, 155
Birth & death process
   birth rate, 220, 221
   death rate, 220, 221
   population model, 217, 220
   pure birth process, 217, 219, 220
Birth & Death (B&D) queueing model
   arrival rate, 221, 233, 234
   balance equations, 217, 233, 235–237, 249
   departure rate, 221, 233
   entering rate, 229, 234
   general birth & death queueing model, 217, 221, 233, 235, 236, 247
   infinite capacity, 221–223, 233, 235, 236, 239, 240
   Kendall's notation, 233
   $M/M/1$ queueing model, 217, 221, 222, 233, 235
   $M/M/1/K$ queueing model, 242, 243, 239–242
   $M/M/m$ queueing model, 217, 223
   state transition diagram, 235, 237, 247, 249
Birthday, 15, 24
Branching process
   probability of eventual extinction, 199, 203, 204, 214
   size of the initial generation, 202
   size of the nth generation, 202
Bridge system, 288
Brownian Motion (BM)
   geometric Brownian motion, 253, 258, 262

© Springer Nature Switzerland AG 2019
E. Bas, *Basics of Probability and Stochastic Processes*,
https://doi.org/10.1007/978-3-030-32323-3

independent increments for Brownian motion, 255
Markovian property, 253, 256, 271
martingale, 265
normal random variable, 253, 255
standard Brownian motion, 253, 257, 261, 265, 271
stationary increments for Brownian motion, 255
stock price, 258, 262
Brownian motion with drift, 253, 257, 258

C

Call center, 87, 88
Central limit theorem, 95, 110, 111, 117, 118, 125, 163, 169, 255
Central limit theorem for a renewal process, 163, 169
Chapman-Kolmogorov equations, 179, 181–183
Characteristic function, 95, 108
Chebyshev's inequality, 110, 111
Coin, 4, 19, 34, 39, 40, 42, 45, 55, 58, 80, 100, 109, 113, 211, 212, 268
Combination
    without sequence consideration, 3, 6, 24
    with sequence consideration, 3, 6, 8
Combinatorics
    combination, 3, 6, 8
    permutation, 3, 5, 8, 9
Communication, 179, 183–188, 191–193, 196–198
Communication classes, 179, 183–188, 191–193
Complementary cumulative distribution function, 39–41, 43, 44, 78
Complement of an event, 18
Compound Poisson process, 149, 156, 160
Conditional distribution, 95, 105
Conditional expected value, 95, 105
Conditional probability, 27, 28, 105, 179, 199, 273
Conditional variance, 95, 105
Conservation equation, 233
Constant Failure Rate (CFR), 275, 276, 289
Continuity correction, 79, 80, 160
Continuous random variable
    exponential random variable, 71, 73–75, 81, 83, 84, 86–89, 91
    gamma random variable, 71, 80, 81, 84
    lognormal random variable, 71, 81, 82, 84

normal random variable, 71, 75–78, 80, 84, 87
standard (unit) normal random variable, 71, 77–80, 84, 90, 92, 93
uniform random variable, 71–73, 84, 85
Weibull random variable, 71, 82–84, 91
Continuous-time Markov chain
    birth & death process, 218–220
    Chapman-Kolmogorov function, 179, 181–183
    infinitesimal generator matrix, 217, 224, 226
    Kolmogorov's backward equation, 225, 230, 231
    Kolmogorov's forward equation, 225, 231
    stationary property, 180, 181
    time-reversibility, 217, 224, 227, 251
    transition probability function, 180–183, 225
Continuous-time stochastic process, 125–127, 129, 135, 156
Convolution
    convolution of continuous random variables, 95, 101
    convolution of discrete random variables, 95, 101
Correlation, 95, 102
Counting process
    independent increments, 134, 136, 137
    Poisson process, 135–137
    renewal process, 134, 135
    stationary increments, 134, 136, 138
Covariance, 95, 102
Cumulative distribution function, 39–43, 59, 77, 85, 92, 96, 100, 116, 165, 273, 274, 276, 287

D

Decreasing Failure Rate (DFR), 275
Density function, 39–41, 43, 47, 71–73, 75–77, 80–83, 85, 95, 96, 98, 101, 102, 105, 114, 116, 121, 122, 149, 156, 165, 273, 274, 276, 287
Die, 3, 4, 15, 16, 19, 27, 46, 92, 102, 118, 120, 221, 227
Discrete random variable
    Bernoulli random variable, 55, 60
    binomial random variable, 55–58, 60, 63, 66, 68
    geometric random variable, 55, 58, 60
    hypergeometric random variable, 55, 59, 61, 65, 66

negative binomial random variable, 55, 57, 58, 60

Poisson random variable, 55–57, 61, 63, 66

Discrete-time Markov chain
 branching process, 199, 202
 Chapman-Kolmogorov function, 179, 181–183
 hidden Markov chain, hidden Markov model, 199, 204, 205, 211, 212
 n-step transition probability, 183
 n-step transition probability matrix, 182, 183
 one-step transition probability, 180, 182
 one-step transition probability matrix, 181, 182
 random walk, 199–201, 214, 215
 rat in the maze, 191, 192
 stationary property, 180, 181
 time-reversible discrete-time Markov chain, 199, 206, 207, 211
 transition probability matrix, 204, 205, 207, 210, 212, 214, 215

Discrete-time stochastic process, 126–128, 204

Distribution function, 42

Doob type martingale, 265, 268

E

Elementary renewal theorem, 163, 165, 166, 170

Embedded discrete-time Markov chain, 180, 183, 186, 187

Emission, 83, 91, 204, 211–213

Emission matrix, 205, 212

Ergodic Markov chain, 179, 190, 197, 198

Ergodic state, 183, 189, 190

Ergodic theorem, 190

Erlang distribution, 235

Event
 complement of an event, 18
 intersection of events, 18
 mutually exclusive event, 12, 15–17, 19, 23, 25, 30, 31, 47
 single event, 15–17
 union of events, 17–19

Expected profit, 49, 52, 53

Expected value
 conditional expected value, 95, 105
 expected value by conditioning, 95, 105

Exponential random variable
 further property, 138, 219, 223, 224, 247

intensity function, 147
memoryless property, 74, 86, 89, 91, 140, 143, 145, 277, 290
rate, 74, 87, 88, 145, 275, 276

F

Failure rate function
 bathtub curve, 275
 constant failure rate (CFR), 275, 276, 289
 decreasing failure rate (DFR), 275
 increasing failure rate (IFR), 275

First passage time, 126, 127

Floor function, 173

Forward recurrence time, 167

Further property, 138, 219, 223, 224, 247

G

Gambler's ruin problem, 199, 200

Gamma function, 84

Gamma random variable
 gamma function, 84

Generalized basic principle of counting, 3, 4, 23

Geometric Brownian motion
 lognormal random variable, 253, 258
 stock price, 258, 262

Geometric random variable, 55, 58–60, 74, 112

H

Hazard rate function, 274

Hidden Markov chain, hidden Markov model
 emission, 204, 211–213
 emission matrix, 205, 212

Hitting time, 125–129

Homogeneous Poisson process
 pure birth process, 217, 219, 220

Hypergeometric random variable, 55, 59, 65, 66

I

IFR distribution, 275

Increasing Failure Rate (IFR), 275

Independence
 independence of three events, 32
 independence of two events, 32

Independent and identically distributed (i.i.d.), 100, 126

Independent increments, 131, 134, 136, 137,
    141, 147, 150–153, 253–256, 262,
    271, 272
Indicator variable, 46
Infinitesimal generator matrix, 217, 224, 226
Inspection paradox, 163, 169
Instantaneous transition rate, 189, 225, 227,
    252
Integrating the tail function method, 41, 44,
    47, 48
Intensity function, 135–137, 146, 150–152
Interarrival time, 131–135, 138, 145, 149,
    163–167, 169, 222, 235, 243
Intersection of events, 18
Irreducible Markov chain, 183, 184, 198
Irwin-Hall distribution, 122

J
Job scheduling problem, 3, 10
Joint cumulative distribution function, 96
Jointly distributed random variables
    convolution, 100, 101
    correlation, 102
    covariance, 102
    joint cumulative distribution function, 96
    joint probability density function, 96, 98,
        101, 102
    joint probability mass function, 96
    order statistics, 100, 101
    sums of independent random variables,
        98, 99
Joint probability density function, 95, 96, 98,
    101, 102, 114, 121, 122, 149, 156
Joint probability density function of order
    statistics, 101, 121, 122, 156
Joint probability mass function, 95, 96

K
Kendall's notation
    Erlang distribution, 235
    general distribution, 235
    general independent distribution, 235
Knapsack problem, 3, 11
Kolmogorov's backward equation
    instantaneous transition rate, 225
Kolmogorov's forward equation
    instantaneous transition rate, 225
Kolmogorov's inequality
    martingales, 269, 270
    submartingales, 265, 270

L
Limiting probability of a state of a Markov
    chain
    balance equations, 189, 190
Limit theorems
    central limit theorem, 95, 110, 111, 117,
        118, 125, 163, 169, 255
    Chebyshev's inequality, 111
    Markov's inequality, 110, 270
    strong law of large numbers, 95, 110,
        111, 125
    weak law of large numbers, 110
Little's law
    average amount of time, 234, 241, 242
    average arrival rate, 234
    average number of objects, 234, 240,
        241, 243
Lognormal random variable
    down time, 82
    fatigue failure, 82
    stock price, 82, 258

M
Markov chain
    absorbing state, 185
    aperiodic state, 187, 197, 198
    Chapman-Kolmogorov equations, 179,
        181–183
    communication, 179, 185, 191, 193,
        196–198
    communication classes, 179, 183–188,
        191–193, 196
    continuous-time Markov chain, 179,
        180, 182–184, 186, 188–190, 217, 218
    discrete-time Markov chain, 179, 180,
        182–187, 199, 205–207, 211
    embedded discrete-time Markov chain,
        179, 180, 183, 185–187, 189, 191, 199
    ergodic Markov chain, 179, 190, 197,
        198
    ergodic state, 183, 189, 190
    irreducible Markov chain, 183, 184, 198
    limiting probability of a state, 179, 189,
        190
    Markovian property, 180, 205, 206, 253,
        256, 271
    period of a state, 179, 183, 187, 188
    reachability, 179, 184, 185, 191–193,
        196–198
    recurrence, 186
    recurrent state, 186, 187, 192, 194, 196,
        197

reducible Markov chain, 192, 197
state space, 180–184, 189, 190, 193, 194, 200, 201, 207, 209, 210, 214, 226
state transition diagram, 179, 184, 185, 188, 191–193, 196–198, 205, 206, 212, 214, 215, 235, 237, 239, 242, 245, 247, 249, 250
steady-state probability, 189
transience, 186
transient rate, 189, 225–227, 252
transient state, 179, 183, 186, 187, 191–194, 196, 197, 201, 218
Markov decision process
randomized policy, 208, 210
Markovian, 180, 205, 206, 222, 235, 253, 256, 271
Markov's inequality, 110, 270
Martingale
Azuma-Hoeffding inequality, 265, 269, 270
Doob type martingale, 265, 268
fair game, 266
Kolmogorov's inequality, 265, 269, 270
martingale convergence theorem, 265, 269, 271
submartingale, 265, 267, 270
supermartingale, 265, 267, 268
Martingale convergence theorem, 265, 269, 271
Mean Down Time (MDT), 273, 278, 279
Mean Residual Life (MRL), 273, 277, 289, 290
Mean Time Between Failures (MTBF), 273, 278, 279
Mean Time To Failure (MTTF), 273, 276, 289, 290
Mean Time To Repair (MTTR), 273, 278, 279, 290
Mean Up Time (MUT), 273, 278, 279
Memoryless property, 74, 75, 86, 89, 91, 140, 143, 145, 222, 277, 290
Minimal cut set, 283, 290, 291
Minimal cut vector, 283
Minimal path set, 283, 290, 291
Minimal path vector, 283
*M/M/*1 queueing model, *M/M/*1 system
balance equations, 233, 235, 236, 239, 249, 250
bulk service, 233, 249, 250
Little's law, 233, 235, 239, 240
state transition diagram, 235, 239, 249
*M/M/*1/*K* queueing model
balance equations, 242, 243

Little's law, 243
state transition diagram, 242
M/M/m queueing model, 217, 223
Moment generating function, 95, 108, 109, 112
Multiplication rule, 29, 30, 199, 212
Mutually exclusive events, 12, 15–17, 19, 23, 25, 30, 31, 47

N
Negative binomial random variable, 55, 57, 58, 60
Negative dependency, 273, 284
Newsboy problem, newsvendor problem, 39, 49
Nonhomogeneous Poisson process, 131, 135, 136, 146, 149–152
Nonrepairable item
time to failure, 274, 276, 286, 289
Nonstationary increments, 136
Normal random variable
68%, 95%, 99.7% rule, 76
continuity correction, 79, 80
normal random variable approximation to binomial random variable, 71, 79, 80, 90, 93, 149, 154
N–step transition probability, 183
N–step transition probability matrix, 182, 183, 194

O
One-step transition probability, 180, 182
One-step transition probability matrix, 181, 182
Order statistics
joint probability density function of order statistics, 101, 121, 122, 156

P
Parallel system, 33, 284, 285
Partitioning a homogeneous Poisson process, 149, 153
Period of a state, 179, 183, 187, 188
Permutation
without repetition, 3, 5, 8, 9
with repetition, 3, 5
Point process, 131–133
Poisson Arrivals See Time Averages (PASTA), 233, 234
Poisson process
arrival time, 138, 149

compound Poisson process, batch Poisson process, 156
homogeneous Poisson process, 131, 135–139, 141, 149–151, 153, 154, 156–158, 160, 173, 174, 217, 219, 220, 253, 271, 272
independent increments, 131, 136, 137, 141, 147, 150–153
intensity function, 135–137, 146, 147, 150–152
interarrival time, 131–135, 138, 145, 149
nonhomogeneous Poisson process, 131, 135, 136, 146, 149–152
nonstationary increments, 136
partitioning, 149, 153, 154
stationary increments, 136, 138, 141, 142, 149, 152, 153, 157, 158, 180, 271, 272
superposition, 149, 154
thinning, 153
waiting time, 138, 233
Poisson random variable
approximation to binomial random variable, 56
typographical errors, 63, 64
Population, 127, 199, 202–204, 213, 214, 217, 220, 221, 228, 229, 235
Positive dependency
square root method, 273, 284, 285
Probability, 3, 15–43, 45–48, 50–52, 55, 59–68, 71–78, 80–83, 85–92, 95–98, 100–103, 105, 107, 110, 111, 113–118, 120–122, 125, 126, 136, 137, 139, 143–147, 149–156, 158–160, 164, 165, 171, 172, 176, 178–183, 185, 186, 188–195, 199–215, 217–225, 227–230, 234, 241, 244, 246, 250–252, 254, 257, 261, 262, 265, 267, 271, 273, 274, 276, 278–282, 285, 287–290
Probability density function, 39–43, 47, 48, 71–73, 75–77, 80–83, 85, 95, 96, 98, 101, 102, 114, 116, 121, 122, 156, 165, 273, 274, 276, 287
Probability mass function, 39–42, 46, 59, 80, 95–98, 120, 164
Probability of eventual extinction, 199, 203, 204, 214
Probability theory, 15, 17, 18
Probability vector, 273, 280–282, 288

Q
Quality

acceptance sampling, 55, 66
Queueing models
average amount of time, 234, 239, 241, 242
average number of objects, 240, 243
conservation equation, 233
Kendall's notation, 233, 235, 237
Little's law, 233, 239, 240, 243
PASTA, 233, 234

R
Random experiment
coin, 4, 113
die, 3, 4, 16, 19
trial, 16
urn, 4, 113
Random observer property, 234
Random variable
complementary cumulative distribution function, 39–41, 43, 44, 78
continuous random variable, 39–45, 47, 48, 71, 73–75, 80, 82–84, 95, 96, 98, 100–103, 105, 108, 109, 114, 122
cumulative distribution function, 39–44, 59, 77, 78, 85, 92, 96, 100, 116, 165, 273, 274
density function, 40–43, 47, 72, 73, 75, 77, 80, 82, 85, 95, 96, 98, 101, 102, 105, 114, 121, 122, 156, 165, 273, 274, 276
discrete random variable, 39–42, 45, 52, 55, 59, 60, 74, 96, 101, 105, 108, 109, 173
distribution function, 39–43, 77, 78, 96, 100, 116, 165, 273, 276, 287
expected value, 39–45, 47, 49, 51, 59–64, 68, 75, 78, 83, 85, 88, 89, 95, 102, 103, 105, 114, 115, 145, 149, 160, 258, 266, 273
indicator variable, 46
integrating the tail function method, 41, 44, 47, 48
probability density function, 39–43, 47, 48, 71–73, 75–77, 80–83, 85, 96, 98, 101, 102, 114, 121, 122, 149, 156, 165, 273, 274, 287
probability mass function, 39–42, 46, 59, 80, 95–98, 120, 164
standard deviation, 39, 40, 42–44, 102, 117
tail function, 39–41, 44, 47, 48, 78
variance, 39, 40, 42–45, 47, 48, 59–64, 68, 75, 78, 83, 85, 86, 88, 89, 95,

103–106, 109, 115, 140, 145, 157, 160, 169, 202–204, 255–259
Random walk
  Gambler's ruin problem, 199, 200
  simple random walk, 199–201, 267, 268, 270
  simple symmetric random walk, 199, 200, 214, 215, 253–255, 266
Rate, 74, 87, 88, 116, 138, 139, 141–143, 145, 146, 150, 151, 153, 154, 156–160, 163, 165, 166, 176, 189, 213, 218–227, 229, 230, 233, 234, 236, 237, 239, 241, 243, 245–250, 252, 271, 273–276, 279, 286, 287, 289, 290
Rat in the maze, 191, 192
Reachability, 179, 184, 185, 191–193, 196–198
Recurrence, 167, 186
Recurrent state
  null recurrent state, 187
  positive recurrent state, 187
Reducible Markov chain, 192, 197
Regenerative process
  alternating renewal process, 163, 170, 171, 178, 273, 279
  renewal process, 131, 134, 135, 145, 150, 151, 163–170, 172, 175, 273
Reliability function, 273–277, 287
Reliability theory
  availability, 273, 274, 279–281
  failure rate function, hazard rate function, 274
  mean residual life (MRL), 273
  mean time to failure (MTTF), 273
  minimal cut set, 283, 290, 291
  minimal cut vector, 283
  minimal path set, 283, 290, 291
  minimal path vector, 283
  nonrepairable item, 273, 274, 276
  reliability function, survival function, 274, 275
  repairable item, 273, 274
  systems with dependent components, 273, 284
  systems with independent components, 273, 280
Renewal equation, 163, 165, 173
Renewal function, 163, 165, 166
Renewal process
  age, 163, 167, 168, 175
  alternating renewal process, 163, 170, 178, 273, 279

average age, 167, 168
average excess, 167, 168, 175
average spread, 167
backwards recurrence time, 167
central limit theorem for a renewal process, 163, 169
elementary renewal theorem, 163, 165, 166, 170
excess, 163, 167, 168, 175
forward recurrence time, 167
inspection paradox, 163, 169
limit theorem, 163, 165, 166, 169, 255
renewal cycle, 164, 166, 167, 177
renewal equation, 163, 165, 173
renewal function, 163, 165, 166
renewal reward process, 163, 165–167, 177
spread, 163, 167, 168, 175
Repair, 74, 85, 86, 164, 166, 176, 233, 247, 248, 273, 278, 279
Repairable item
  mean down time (MDT), 273, 278, 279
  mean time between failures (MTBF), 273, 278, 279
  mean time to repair (MTTR), 273, 278, 279, 290
  mean up time (MUT), 273, 278
Roulette wheel, 66

S
Sample space, 15, 16, 18, 19, 28, 39, 40, 66, 195
Server, 88, 131, 142–144, 217, 222, 224, 229, 233, 235, 237, 241, 245, 249, 250
Set theory
  Venn diagram, 15, 20, 21, 25, 34
Simple random walk
  Gambler's ruin problem, 199, 200
Simple symmetric random walk
  extension to a simple symmetric random walk, 253–255
  restricted state space, 214
Single event, 15–17
Size of the initial generation, 202
Size of the nth generation, 202
Square root method, 273, 284, 285
Standard Brownian motion, 253, 257, 261, 265, 271
Standard deviation, 39, 40, 42–44, 102, 117
Standard (unit) normal random variable, 71, 77–80, 84, 90, 92, 93

State space, 125, 126, 128, 129, 135,
    180–184, 189, 190, 193, 194, 200,
    201, 207, 209, 210, 214, 226
State transition diagram, 179, 184, 185, 188,
    191–193, 196–198, 205, 206, 212,
    214, 215, 235, 237, 239, 242, 245,
    247, 249, 250
State vector, 273, 280, 281, 283
Stationary increments, 134, 136, 138, 141,
    142, 149, 152, 153, 157, 158, 180,
    255, 256, 260, 262, 271, 272
Stationary property, 180, 181
Steady-state probability, 189
Stochastic process
    continuous-time stochastic process,
        125–127, 129, 135, 156
    counting process, 131–135, 163, 164
    discrete-time stochastic process,
        126–128, 204
    first passage time, 126, 127
    hitting time, 125–129
    point process, 131–133
    Poisson process, 131–133, 135–139,
        141, 146, 149–154, 156–160
    state space, 125, 126, 128, 129
    stopping time, 125–127
Stock price, 37, 38, 82, 127, 258, 262
Stopping time, 125–127
Straight in a poker hand, 15, 24
Strong law of large numbers, 95, 110, 111,
    125
Structure
    bridge system, 288
Structure function
    k-out of-n structure, 281
    parallel structure, 281, 282
    series structure, 281, 282
Submartingale
    superfair game, 267
Sums of independent random variables
    sum of independent binomial random
        variables, 99, 100, 113
    sum of independent exponential random
        variables, 99
    sum of independent normal random vari-
        ables, 145
    sum of independent Poisson random
        variables, 99, 114, 145, 172
    sum of independent uniform random
        variables, 99
Supermartingale
    subfair game, 267

Superposition of a homogeneous Poisson
    process, 154
Survival function, 274, 275
Systems with dependent components
    β-factor model, 273, 286
    negative dependency, 273, 284
    positive dependency, 273, 284
    square root method, 273, 284, 285
Systems with independent components
    availability function, 280, 282, 283
    minimal cut set, 283
    minimal cut vector, 283
    minimal path set, 283
    minimal path vector, 283
    probability vector, 280–282, 288
    state vector, 273, 280, 281, 283
    structure function, 273, 280–282

T
Tail function, 39–41, 44, 47, 48, 78
Thinning a homogeneous Poisson process,
    153
Time-reversibility, 217, 224, 227, 251
Time-reversible Markov chain
    reversed process, 206
Time to failure, 273–277, 286, 289, 290
Total probability formula, 30
Transience, 186
Transient state, 179, 184, 186, 187, 191–194,
    196, 197, 201, 218
Transition probability matrix, 181–183, 185,
    186, 188, 190, 191, 193–195, 204,
    205, 207, 210, 212, 214, 215, 218
Transition rate
    instantaneous transition rate, 189, 225,
        227, 252
Traveling salesperson (salesman) problem,
    3, 9
Tree diagram, 27, 29–31, 35–38
Trial, 16, 55–59, 61, 62, 80, 111, 154, 155
Triangular random variable, 99, 121
Typographical errors, 44, 45, 55, 63, 64, 117,
    128, 129

U
Uniform random variable
    Irwin-Hall distribution, 122
    triangular random variable, 99, 121
Union of events
    union of three events, 18
    union of two events, 18
Universal set, 18

Urn, 4, 21, 29, 35, 49, 55, 59, 62, 64, 96, 103, 105, 107, 113, 128

V

Variance
   conditional variance, 95, 105
   variance by conditioning, 95, 105
Venn diagram, 15, 20, 21, 25, 34

W

Waiting time, 72, 74–76, 78, 132, 138, 143, 233
Weak law of large numbers, 110
Weibull random variable
   scale parameter, 82
   shape parameter, 82
Without replacement, 29, 55, 59, 62, 66
With replacement, 66, 113, 128

Printed in the United States
By Bookmasters